Biodiversity of the Southern Ocean

Series Editor
Françoise Gaill

Biodiversity of the Southern Ocean

Bruno David
Thomas Saucède

First published 2015 in Great Britain and the United States by ISTE Press Ltd and Elsevier Ltd

Apart from any fair dealing for the purposes of research or private study, or criticism or review, as permitted under the Copyright, Designs and Patents Act 1988, this publication may only be reproduced, stored or transmitted, in any form or by any means, with the prior permission in writing of the publishers, or in the case of reprographic reproduction in accordance with the terms and licenses issued by the CLA. Enquiries concerning reproduction outside these terms should be sent to the publishers at the undermentioned address:

ISTE Press Ltd
27-37 St George's Road
London SW19 4EU
UK

www.iste.co.uk

Elsevier Ltd
The Boulevard, Langford Lane
Kidlington, Oxford, OX5 1GB
UK

www.elsevier.com

Notices
Knowledge and best practice in this field are constantly changing. As new research and experience broaden our understanding, changes in research methods, professional practices, or medical treatment may become necessary.

Practitioners and researchers must always rely on their own experience and knowledge in evaluating and using any information, methods, compounds, or experiments described herein. In using such information or methods they should be mindful of their own safety and the safety of others, including parties for whom they have a professional responsibility.

To the fullest extent of the law, neither the Publisher nor the authors, contributors, or editors, assume any liability for any injury and/or damage to persons or property as a matter of products liability, negligence or otherwise, or from any use or operation of any methods, products, instructions, or ideas contained in the material herein.

For information on all our publications visit our website at http://store.elsevier.com/

© ISTE Press Ltd 2015
The rights of Bruno David and Thomas Saucède to be identified as the authors of this work have been asserted by them in accordance with the Copyright, Designs and Patents Act 1988.

British Library Cataloguing-in-Publication Data
A CIP record for this book is available from the British Library
Library of Congress Cataloging in Publication Data
A catalog record for this book is available from the Library of Congress
ISBN 978-1-78548-047-8

Contents

Preface ix

Introduction xi

Chapter 1. A Brief History of Exploration and Discovery 1

    1.1. The Age of Navigation 1
    1.2. Scientific expeditions come to the fore 6
        1.2.1. Pre-1914, the precursor era 6
        1.2.2. Post 1950, the age of permanent settlement 9
    1.3. An increase in commercial exploitation 9
    1.4. Dynamics of the discovery of Southern Ocean biodiversity 12
    1.5. Tools for oceanography exploration 13

Chapter 2. The Southern Ocean and its Environment: a World of Extremes 17

    2.1. An ocean with undefined limits 18
    2.2. The southern climate: windy and cold, with very little light 19
        2.2.1. Strong winds. 19
        2.2.2. Extreme cold. 20
        2.2.3. From winter night to weak daylight 21
    2.3. Ice in all its forms 22
        2.3.1. Sea ice 22
        2.3.2. Ice sheets and ice shelves 23
    2.4. In isolation yet interconnected, the complexity of ocean circulation . . 24
        2.4.1. Ocean currents 24
        2.4.2. Ocean fronts and the zonation of water masses 26
        2.4.3. A complex interplay between wind, water and ice 26
        2.4.4. Water in action 28

2.5. Sediment and nutrients 28
   2.5.1. Marine sediment and its origins 28
   2.5.2. Oxygen and nutrients, sources of marine life 29

Chapter 3. The Ocean Through Time 33

3.1. The split of a supercontinent from the Jurassic to the Eocene 34
3.2. Global cooling at the Eocene-Oligocene transition 38
3.3. Other thermal anomalies during the Oligocene and Miocene 40
3.4. Another cold snap in the late Miocene 40
3.5. Climatic oscillations and glacial-interglacial cycles 41

Chapter 4. Southern Ocean Biogeography and Communities 43

4.1. Inventorying Antarctic marine biodiversity 44
4.2. Southern Ocean biogeography 45
   4.2.1. A rich ocean 45
   4.2.2. Unique biodiversity 45
   4.2.3. Richness and latitude 47
   4.2.4. Biogeographic regions and provinces 49
   4.2.5. The paradox of bipolar distribution patterns 56

Chapter 5. History of Biodiversity in the Southern Ocean 59

5.1. So much ice yet so few fossils 60
5.2. Origins and age of Antarctic marine biodiversity 61
5.3. Break-up of Gondwana and isolation of Antarctic fauna 63
5.4. Mass extinction event at the end of the Mesozoic Era 64
5.5. Evolution of biodiversity and ancient climatic changes 65
   5.5.1. The Paleocene-Eocene Thermal Maximum 65
   5.5.2. Consequences of the late Eocene biological crisis 65
   5.5.3. Glaciation and species adaptation in the Miocene Epoch 67
   5.5.4. Are glacial-interglacial cycles good for biodiversity? 68

Chapter 6. Adaptation of Organisms 71

6.1. Surviving the cold and escaping the ice 71
   6.1.1. Fish that make their own antifreeze 71
   6.1.2. Looking out for number one, but stronger together 73
   6.1.3. A good insulator 73
   6.1.4. Adaptations in physiology and metabolism 74
6.2. Living with ice 75
   6.2.1. Sea ice habitats 75
   6.2.2. Far from the world, under the ice shelves 77
6.3. Dealing with intense fluctuations 79

  6.3.1. Hellish coastline conditions. 79
  6.3.2. Advantaged trophic groups 79
  6.3.3. Feeding their young by endless periods of fasting. 81
  6.3.4. From total night to permanent day 81
 6.4. Lower metabolic rates, longer lifespans and gigantism 81
  6.4.1. Metabolism and development. 82
  6.4.2. Long-lived forms 84
  6.4.3. Gigantism 85
 6.5. Parents caring for their offspring 87
  6.5.1. Two strategies 87
  6.5.2. Kangaroo sea urchins 89
  6.5.3. Why is there so much brooding in the Southern Ocean? 90

Chapter 7. Projections into the Future 93

 7.1. The immediate future 93
  7.1.1. Invasive species 95
  7.1.2. Extinctions 97
  7.1.3. Acidification 99
 7.2. The next cold event 100
 7.3. Drifting continents 101

Appendix 103

Bibliography 105

Index 115

Preface

This book is intended for readers curious to learn more about the vast ocean surrounding the most isolated and inhospitable continent on the planet, the Southern Ocean. This ocean, which is isolated geographically (in a near polar position at the end of the Earth), climatically (icy cold and prone to violent storms) and oceanographically (currents and fronts between water masses form invisible yet effective borders), has very distinct characteristics, as the result of a history that began around 150 million years ago. These characteristics are not without consequences for the evolution and ecology of the biodiversity it supports, a much richer and more bountiful biodiversity than initially expected. Today, this southern biodiversity, untouched by human presence for so long, must in its turn face global climate change. Unable to migrate further south, how will it react and cope? These are the unique aspects that we have chosen to highlight in this book.

The writing of this book owes much to our frequent visits to southern regions and to the exciting expeditions we have taken part in. Research missions to the Southern Ocean are, by their very nature, and of necessity, collaborative adventures. This work has drawn unreservedly on the lessons learned and the experience gained over time through contact with our colleagues and all those who participated in these scientific missions. Some will undoubtedly recognize themselves in the following pages. Such missions are only possible because major projects allow significant infrastructure to be made accessible to all researchers. We would like to thank the French Polar Institute (IPEV), the French Southern and Antarctic Lands (TAAF), the Alfred Wegener Institute (AWI), the British Antarctic Survey (BAS), the Chilean Antarctic Institute (INACH) and the Scientific Committee on Antarctic Research (SCAR).

Finally, we would like to express our deepest gratitude to Clément Blondel, Gilles Boeuf, Carmela Chateau, John Dolan, Guillaume Lecointre, Yvon Le Maho, Rich Mooi, Sébastien Motreuil, Pascal Neige and Catherine Ozouf-Costaz for their support during the writing of this book.

Bruno DAVID
Thomas SAUCÈDE
Dijon
August 2015

Introduction

Antarctica is a vast white desert. Completely inhospitable, it is devoid of almost all life. Microorganisms alone live in the ice, while discrete plant forms, such as mosses and lichens, cling to the few available rocky surfaces. On the fringes of the continent, there are some impressive, albeit rather scattered, penguin colonies (see Figure I.1). Some other birds and a few marine mammals have also found shelter there. In contrast to this icy solitude, the marine waters bordering the continent host a rich and abundant biodiversity. The Southern Ocean teems with marine life, hidden beneath the surface, colonizing the seabed or living under the ice, while billions of organisms thrive in the water column. The marine ecosystems of the Southern Ocean, so difficult to explore because of its remoteness and extreme climate, are amongst the Earth's most unusual yet least known ecosystems. What is known about the evolutionary history of this southern biodiversity? How did it originate and develop? What are its characteristics? What factors govern its distribution and abundance? Research in this area has progressed significantly over recent years, driven by new investigative methods, improved sensor equipment and stronger modeling approaches, which provide ever more pertinent answers to such questions (Figure I.2).

The Southern Ocean (sometimes incorrectly called the Antarctic Ocean) surrounds the entire Antarctic continent and is over 15,000 kilometers in circumference. Greater in length than from Norway to the Cape of Good Hope in the Atlantic, it covers 35 million square kilometers (see Figure A.1 in the Appendix for a map of the region, with the names of the key geographical areas). Isolated in the south of the planet by a large circular current, wracked by almost permanent storms and mostly covered by ice, it shelters an abundance of life with singular characteristics: marked endemism, organisms with slow metabolisms, some with longer lifespans, gigantism, absence of larval developmental stages, etc. These many distinctive features make this ocean a remarkable natural "laboratory", where adaptive, evolutionary and ecological processes can be seen at work, in unusual environmental conditions, rarely found in the past 500 million years of Earth's history.

Figure I.1. *King penguins on Crozet Island © Bruno David*

Long undiscovered, and later imagined as a *Terra australis nondum cognita* or a *Mare glacialis* on Renaissance or 17th and 18th Century maps, the Southern Ocean has been accessible, for both scientific exploration and the exploitation of its fishing resources, for a little more than a century. Recently, it has even become the site of sporting exploits and is starting to attract wealthy tourists. This change means that it is now subjected to anthropogenic pressure, admittedly lower than that affecting other seas, but which could nevertheless upset the balance of its unique ecosystems, already suffering from the global effects of climate change. How will these ecosystems reach a new equilibrium or, conversely, prove their resilience? What can be said of their possible future and how should we address this issue?

Recent efforts in the exploration and study of Antarctic marine biodiversity have greatly benefited from a highly favorable context of international collaboration. The

International Polar Year, from 2007 to 2008, and the CAML (Census of Antarctic Marine Life) inventory project, between 2005 and 2010, have provided a unique example of large-scale collaboration between biologists. The International Polar Year brought together over 300 scientists from 37 different countries and helped to coordinate campaigns for 18 oceanographic research vessels. By compiling all the data collected and providing access via database networks like the SCAR *Marine Biodiversity Network* (www.biodiversity.aq), it has been possible to summarize our knowledge of Antarctic biodiversity by integrating all previous information. In 2014, this summary led to the publication of an atlas of Antarctic biodiversity, the *Biogeographic Atlas of the Southern Ocean* [DEB 14], one version of which can be accessed online at the above-mentioned website. This atlas combines the work of 140 scientists and brings together over a million items of biogeographic data pertaining to 9,000 species collected from 400,000 different sites.

Figure I.2. *Divers inventorying the coastal marine biodiversity of the Kerguelen Islands* © Sébastien Motreuil

This book is divided into seven chapters. After this brief introduction, the first chapter will trace the history of expeditions and discoveries, from the first human incursions to the most recent scientific investigations, as well as the resources allotted to them. The second chapter focuses on defining the Southern Ocean. Why differentiate it from the southern parts of other oceans? What are its limits? What are its main features? The third chapter discusses the geological and climatic history of the region, beginning with the break-up of the supercontinent, Gondwana, 150 million years ago, leaving one fragment, the Antarctic continent, isolated in a polar position, surrounded by a huge ocean. The next three chapters form the heart of the book, addressing many questions in relation to the Southern Ocean's biodiversity. These include its history, concomitant with the global cooling that has occurred over the past 40 million years, as shown by the fossil record; its evolutionary characteristics (why are certain groups more successful here than others?); the factors influencing its distribution (does regionalism exist within this ocean?); the relationship between southern biological communities and those in other oceans (what exchanges and what barriers are there?); and this biodiversity's ability to adapt to extreme conditions (what solutions does it find to survive?). Finally, Chapter 7 deals with the outlook for the Southern Ocean, not only in the short term, but also many tens or hundreds of millennia in the future.

In these, the coldest waters on the planet, researchers can discover and observe situations unequalled anywhere else on Earth.

1

A Brief History of Exploration and Discovery

It is now acknowledged that *Homo sapiens* settled in the farthest reaches of South America a very long time ago [GOE 08, MIS 12]. Their arrival dates back tens of thousands of years and might even have preceded their installation in North America. How did those pioneer settlers travel so far, so long ago? This question has not yet been answered, but the first humans to venture through the channels in the region of Patagonia, Tierra del Fuego, and as far as Cape Horn were the ancestors of the Alacalufe, the Ona, the Yaghan or the Haush. They were the contemporaries of the Mylodon, the giant sloth of the Pleistocene Epoch, which remains a source of pride for the little southern town of Puerto Natales, where a life-sized statue of a Mylodon stands on a roundabout. It was not until Ferdinand Magellan's expedition in 1520 that European settlers first reached these distant lands. In 1616, the route via Cape Horn was discovered by Dutch sailors from the town of Hoorn, where merchants were seeking to avoid the Dutch East India Company's monopoly of the Strait of Magellan. This expedition did not exactly "round the Horn", as they sailed slightly north of the Horn archipelago, but it provided the first glimpse of the harshness of the as yet totally mysterious Southern Ocean. The exploration of this region of the world lasted for several centuries; it was described as *Terra Australis nondum cognita* by the Belgian cartographer, Abraham Ortelius, in 1570, thus hinting at hopes of land to the south, while mapping a completely imaginary contour.

1.1. The Age of Navigation

In 1578, while sailing in the region, Sir Francis Drake had been blown off course towards the Diego Ramirez Islands (56° S), thus giving his name to the 700-kilometer passage separating the tip of South America from the Antarctic

Peninsula. However, the first to deliberately set sail for the south and doubtless the first to reach the archipelagos surrounding the peninsula must have been whalers or seal hunters, who have left no recorded trace in maritime history. It was not until the 17th and 18th Centuries that navigators commissioned by their sovereigns deliberately sought to sail as far south as possible in the hope of finding that *Terra Australis* represented on certain maps as far larger, and therefore far more promising, than it eventually turned out to be. In 1675, the English mariner Antonio de la Roché was probably the first to lay eyes on South Georgia (54° S). Much further east, in 1739, whilst sailing under the French flag for Louis XV, Jean-Baptiste Charles Bouvet de Lozier discovered an island that now bears his name, which has belonged to Norway since 1927 (54° S). Marc-Joseph Marion Dufresne consecutively discovered the Marion and Prince Edward Islands (1771) and the archipelago of Crozet (1772), when he was heading towards the Pacific under orders to take possession of new southern territories. On completing this mission, he left a scroll bearing the King of France's coat of arms on the island known today as Possession Island. He was killed a few months later by Maoris in New Zealand.

At almost the same time, Yves-Joseph de Kerguelen Trémarec approached the Duke of Praslin, naval minister under Louis XV, to suggest an expedition to the south of the Indian Ocean with the intention of adding to French possessions such as Bourbon Island (Reunion Island, today) and Isle of France (Mauritius, today). Embarking in 1771 on the corvette *La Fortune* accompanied by the store-ship *Le Gros Ventre*, he discovered a land of wilderness at 49° S and named it "Austral France" (later called Desolation Island by James Cook, and today known as the Kerguelen Islands). On his return to France, having lost sight of *Le Gros Ventre* during a storm, he embellished his description of the islands to the king, extolling their size and potential resources and describing them as a vast continent. Covered with honors and commended as a hero, Kerguelen obtained the king's approval for a second expedition to the same region. He departed from Brest in 1773 for this second journey with two ships carrying future colonists. Having smuggled his young mistress aboard disguised as a valet, he was faced with a cold welcome in Mauritius and quickly set sail for the Kerguelen Islands. On reaching the archipelago, the expedition barely set foot on the island and not a single colonist disembarked. The weather was terrible and Kerguelen refused to delay and explore the islands. The crew attributed this complete failure to the presence of a woman onboard. Back in Paris, with a new king on the throne, Kerguelen was criticized for his actions, notably his abandonment of *Le Gros Ventre* (which returned to France much later after having endured a long journey), his lies and the failure of his expedition. He was sentenced to prison and incarcerated. On his release in 1778, he joined the revolution and died in 1797.

In 1810, Frederick Hasselborough discovered Macquarie Island, south of New Zealand (54° S). In 1819, William Smith landed on King George Island in the South Shetland Islands, close to the Antarctic Peninsula (62° S). In 1823, James Weddell reached 74° S in the sea that today bears his name.

Meanwhile, James Cook, a contemporary of Marion Dufresne and Kerguelen, and a talented opponent in the fierce rivalry between the English and the French in their respective bids to obtain southern lands, had previously completed the incredible feat of passing through the Antarctic Circle in January 1773, onboard the HMS Resolution. It was only bad weather and bad luck that prevented him from finding the Antarctic continent during that voyage. The controversy over who first found the continent still persists: the Englishmen William Smith and Edward Bransfield in 1819 and 1820, the Russian Fabian Gottlieb von Bellingshausen or the American Nathaniel Palmer in 1820-1821? It is possible that the explorer and seal hunter John Davis first set foot on the Antarctic Peninsula in 1821, but the Frenchman Jules-Sébastien Dumont d'Urville is generally considered to be the first man to have landed on the continent. After leaving Toulon in September 1837, with two corvettes, *L'Astrolabe* and *La Zélée*, following the route of La Pérouse, he began a tortuous trip around the world that eventually, in 1838, brought him close to the Antarctic Peninsula, where he gave the name "Louis-Philippe Land" to the coastline. Two years later, still circumnavigating the world, he decided to leave Hobart (Tasmania) on a second attempt to explore the far south. On January 1st, his two corvettes were heading due south; by January 20th, an icy coastline was in sight and, on January 22nd, two boats landed a handful of men in a small rocky area that they promptly laid claim to and where they collected scientific samples. Dumont d'Urville named this landing site Adelie Land in honor of his wife, Adèle. On February 17th, he returned to Hobart. Today, Adelie Land is the headquarters for the French Antarctic base camp named after Dumont d'Urville, thus reuniting in name the couple, who died together in 1840, in the first French railway disaster. During that southern summer of 1840, competition was high. Towards the end of January, a few days after Dumont d'Urville, the American Charles Wilkes spotted a coastline several hundred kilometers further west and followed it for over 2000 kilometers (today it is called Wilkes Land). On January 30th, one of Wilkes' ships actually crossed paths with *L'Astrolabe*. Over the course of the following southern summer, James Clark Ross's ships (the *Erebus* and the *Terror*) deviated from the routes used by Wilkes and Dumont d'Urville, descended into a deep bay on the continent (the Ross Sea), passed the 78° S mark (close to the McMurdo Sound), skirted the coast of Victoria Land and discovered two huge volcanoes of over 3000 m in height, which he named after the two ships.

Despite these colossal efforts and risks, the Antarctic continent remained poorly known at the end of the 19th Century. On British maps from 1894, only three areas were represented, in a very rough fashion: Graham Land in the Peninsula area,

Wilkes Land and Adelie Land to the south of Australia, and Queen Victoria Land in the Ross Sea. These three areas were sufficient to provide a glimpse of the tremendous size of the southern continent. All that was left was to draw the outline of the coast, a very difficult task owing to the near permanent presence of pack ice and ice floes, making it impossible to distinguish the shore itself. Many islands were therefore mistaken for peninsulas and vice versa. Even icebergs were sometimes confused with islands. In 1914, although the South Pole had been reached by Amundsen two years previously, mapping of the area had barely progressed in 20 years. It was not until modern technological developments (planes, sonar, radar and satellites) that the means existed to map the whole of this 14 million km² continent (see Figure 1.1).

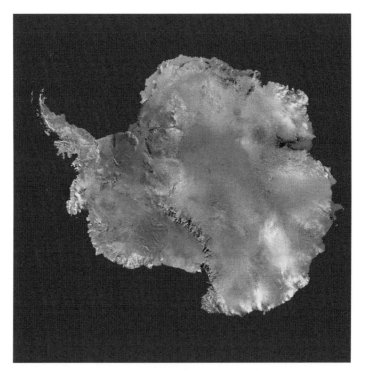

Figure 1.1. *Satellite map of the Antarctic continent (1/80,000,000)* © *NASA*

Shortly before World War II, the Germans were mounting a novel marine and aerial expedition in an unknown area: the region of Queen Maud Land. After conducting an aerial reconnaissance survey of 600,000 km², in January 1939 they claimed the land in the name of the Third Reich and called it Neuschwabenland (New Swabia). Such a claim was not purely for show, but was in fact part of a highly strategic operation whose aim was to claim an area not yet under British control for the German whaling fleet [LÜD 12], in the knowledge that whale oil provided the best source of glycerin for explosives. Just after World War II, the American *Operation Highjump*, technically similar to the 1939 German expedition, collected over 70,000 aerial views. Between 1970 and 1990, satellite technology provided an extremely precise outline of this no longer unknown continent, allowing its exact topography to be established. More recently, seismic imaging has provided a means of seeing through the ice to draw a topographical "beneath the ice" map of the Antarctic, revealing the jagged outline of the land (see Figure 1.2).

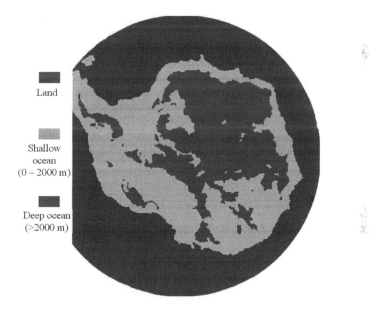

Figure 1.2. *Subglacial map of the Antarctic continent (redrawn from [LYT 01])*

As knowledge increased and became more precise, and as more means of access developed, the necessity of a legal framework became ever more obvious. After much debate, and in contrast with what had happened elsewhere on the planet, scientific interest finally prevailed over economic and political issues. The Antarctic Treaty, signed in Washington in 1959, entered into force from 1961. In 1991, it was reinforced by a protocol on environmental protection (the Madrid Protocol), which stipulates that the Antarctic is an "international natural reserve, devoted to peace and science". This protocol first came into effect in 1998, and has now been ratified by 32 nations.

1.2. Scientific expeditions come to the fore

The first expeditions, such as those of Cook, Dumont d'Urville, Wilkes or Ross, sometimes included a scientific component, as mentioned above, but their main aims were to discover new territories and lay claim to them. Towards the end of the 19th Century, however, expeditions were mounted for purely scientific reasons.

1.2.1. Pre-1914, the precursor era

The first expedition to sail the Southern Ocean in order to take bathymetric soundings and trawl its depths to inventory the fauna was the British HMS Challenger. Early in 1874, during the first round-the-world voyage devoted to oceanography (1873-76), the Challenger explored these cold waters between the 40th and 60th parallels, revealing exotic species of fauna from the far south to the astonished scholars of the time (see Figure 1.3). Some areas of the South Indian Ocean surveyed during this pioneer expedition have not been further explored since that time.

Figure 1.3. *Scientists discover deep-sea fauna. Arrival on the Challenger's deck. Photograph from a Wyville Thomson plate © Bruno David*

Almost simultaneously, the German Baron Georg von Schleinitz's *Gazelle* was exploring the waters around Bouvet Island, the Kerguelen Islands and the region around Enderby Land (1874-76). After the *Challenger* and the *Gazelle*'s pioneering voyages, the next oceanographic expeditions came in two successive waves, interrupted by the Great Depression and the two world wars.

In the 20 years preceding World War I, most expeditions were organized by European countries and were nearly all in the Antarctic Peninsula or the Ross Sea. Foremost amongst these countries, England launched no fewer than five major Antarctic expeditions between 1898 and 1913. The first was the *Southern Cross* (1898-1900), led by the Norwegian Carsten Borchgrevink, followed by the *Discovery* led by Robert Falcon Scott (1901-04), and the *Nimrod* led by Ernest Shackleton (1907-09). Although the next two expeditions had little maritime impact, they nevertheless made history. The fourth expedition, which set sail on the *Terra Nova* in 1910, ended with the tragic death of Scott and his men. The fifth expedition has gone down in history as an epic tale. The *Endurance* drifted and was destroyed by sea ice, but the crew was saved by Ernest Shackleton, who managed to bring them safely to Elephant Island, from where he headed to South Georgia to seek help. To this list could be added a sixth, the Scottish national expedition, privately funded and organized by William Speirs Bruce, which brought back a rich harvest of findings, despite being poorly regarded by the Royal Geographical Society, which saw it as competition for their purely English expeditions.

The Belgian Adrien de Gerlache was the first to endure a winter in the Antarctic, onboard the *Belgica* (during the southern winter of 1898). This heroic feat could have ended in tragedy when the crew almost died of exhaustion after digging a long canal to allow the ship to escape deadly encounters with ice packs. The *Belgica* campaign was innovative in that it was truly international, as de Gerlache had recruited the Norwegian Amundsen, the American Cook, the Hungarian Racovitza, and the Poles Arctowski and Dobrowolski.

The Germans organized several major expeditions: the Hamburger Magellanische Sammelreise to the tip of South America in 1892-93, then the Deutsche Tiefsee on the steamer *Valdivia* (1898-99) and the Deutsche Südpolar on the *Gauss* (1901-03), with scientific teams led respectively by Carl Chun and Erich von Drygalski (see Figure 1.4). Later, Wilhelm Filchner would extensively explore the Weddell Sea (1911) onboard the *Deutschland*.

Two Swedes, navigator Carl Larsen and geologist Otto Nordenskjöld, led the Swedish polar expedition (1901-03). Nordenskjöld wintered on Snow Hill Island with five companions. The following summer, the *Antarctic* was trapped and crushed in the ice before reaching Snow Hill Island, forcing Larsen and his crew to settle on an islet, Paulet (see Figure 7.1), while Nordenskjöld spent a second winter on Snow Hill. They were all eventually rescued by an Argentinian ship in 1903.

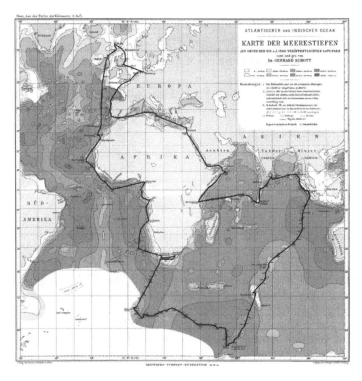

Figure 1.4. *Itinerary of the Valdivia during the Deutsche Tiefsee (1898-99). The Antarctic continent was still largely unknown at the time*

The French, who launched their scientific campaigns much later, benefitted from Jean-Baptiste Charcot's passion for the polar seas. He wintered twice close to the Peninsula, first onboard *Le Français* (1903-05), then on the famous *Pourquoi pas?* (1908-10). Charcot later came to a tragic end onboard the *Pourquoi pas?* in 1936, near Iceland.

In addition to these European contributions, the Australian expedition of the *Aurora* led by Douglas Mawson traced a path around Antarctic regions south of Australia (1911-14). Between 1910 and 1912, the Japanese Shirase Nobu led two campaigns to the Ross Sea aboard the *Kainan Maru*, a small 30 m boat considered laughable by Australians and New Zealanders. On his second campaign, he reached 80° S and was welcomed as a hero on his return to Yokohama.

In summary, the expeditions launched in this era that contributed most to our knowledge of the Southern Ocean and its biodiversity were that of the pioneering *Challenger*, followed by those of the *Gauss* and of the *Valdivia* (Germany), the *Aurora* (Australia) and the *Pourquoi pas*? (France).

1.2.2. Post 1950, the age of permanent settlement

During the time between the two world wars, few large expeditions were launched and it was not until 1950 that scientific activity began again in these remote southern regions. Driven by geopolitical interest, several countries simultaneously decided to build scientific research bases on sub-Antarctic and Antarctic islands, then on the borders of the continent and finally on the ice sheet itself. Keeping the bases stocked required a rotation of ships that were also being used for oceanography research (*Marion-Dufresne, Astrolabe, Aurora Australis, Almirate Oscar Viel*, etc.). Other ships were specifically designed to explore the polar seas, the most emblematic being the Alfred Wegener Institute's F/S *Polarstern* (Germany).

The greatest contributions, doubtless because of the Cold War, came from the United States and the Soviet Union, but also from Germany and France. Other countries making significant contributions include Great Britain, Chile, Argentina, Brazil, Australia, Belgium and, more recently, China. This activity has resulted in the establishment of a multitude of research bases, some only separated by a few hundred meters, as on King George Island, highlighting the political logic underpinning their installation.

The geographical inequality that existed before 1914 has been perpetuated in this latest phase. The main bases were set up in the Peninsula area, and marine research explored both the Weddell and Ross seas as well as the sub-Antarctic Weddell Quadrant (Atlantic Ocean) and Enderby Quadrant (Indian Ocean). A large zone of Antarctic waters to the south of the Pacific Ocean still remains unexplored, notably between the Eights Coast and the King Edward VII Peninsula.

1.3. An increase in commercial exploitation

Quite quickly, geographical exploration and scientific discoveries paved the way for commercial exploitation that became all the more intense as marine resources in the Northern Hemisphere dwindled. In comparison, the untouched southern waters appeared to be incredibly rich in resources. With the 19th Century came a great overexploitation of certain marine resources, with two targets being the most affected.

Whales have historically been represented as an optimal resource, mostly for their meat but more anecdotally for the use of their baleen in the production of umbrellas, corsets and combs. Before the arrival of the gas industry, whale oil was used for lighting. As a result, nearly 200,000 whales were killed between 1804 and 1876 by the whaling fleet of the United States alone. The exceptional quality of whale oil then made it a prime candidate in the production of explosive devices. Colonial conflicts and the arrival of the First World War had a deadly impact on whales in the Southern Ocean: between 1914 and 1917, 175,000 whales were killed around South Georgia. Whale oil retains its lubricant properties even under high pressure. Consequently, it was used in airplane motors and jet engines until synthetic oils and jojoba vegetable oil extract replaced it, as more and more planes were produced. All whale species were affected, resulting in the near disappearance of whales in the Southern Ocean in 1975. Only 360 blue whales, the largest animal on Earth, remained, compared to approximately 240,000 blue whales a mere 70 years earlier. Even today, Japanese "science" culls several hundred southern whales per year as a means of obtaining this commercial resource under the cover of science. However, this situation should improve, as the JARPA II whale-hunting program was pronounced illegal in March 2014 by the International Court of Justice in The Hague. The sustained commercial hunting in the Northern Hemisphere accounts for around 1,100 catches per year, but they do not touch the same populations, despite belonging to the same species. Since the introduction of a ban in 1986, whale populations in the Southern Ocean have been rising rapidly. The total number of individuals belonging to large species now stands at 7,000 (the majority of them being humpback whales) to which number around 500,000 smaller whales can be added (Minke whales).

Fur seals are another example of the violent effects of human pressure. The descriptions provided by James Cook and other 18th Century sailors triggered a rush to find the fur seals of the sub-Antarctic islands and the wild Patagonian coasts. Incidentally, it was one such seal hunter, William Smith, who discovered Livingstone Island on February 19th, 1819 after being blown south during a storm. A new Eldorado was open to these hunters. The first boat to profit from this was undoubtedly the *San Juan Nepomuceno* that unloaded 14,600 skins in Buenos Aires on February 22nd, 1820 [SMI 87]. It is estimated that the following summer (1820-21), during the three-month campaign, between 250,000 and 500,000 fur seals were killed, followed by 400,000 in 1821-22. As recorded in a logbook found on the *Hero*, a small ship belonging to Nathaniel Palmer, who transported the skins from Livingstone beach to transport ships, the daily number of slaughters is shocking: 906 skins on December 5th, 1820, 9700 on December 9th, 5616 on December 12th, 6865 on December 16th, 8229 on December 19th, 8000 on December 30th – three years later the resources were all but exhausted and so the hunters left. After a renewal of hunting by Americans in the 1870s, the rare surviving fur seals were left in peace. The culling was such that it took almost 90 years for small colonies of several dozen animals to be seen again (1958) and then another 30 years for their population to grow significantly (see Figure 1.5).

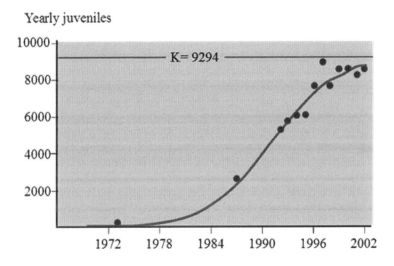

Figure 1.5. *Graph showing the increase in number of young fur seals born per year in two sites in South Shetland. This model (dotted curve) predicts a stabilization of births when population reaches just over 9000 individuals. Modified from [HUC 04]*

Today, the exploitation of fishing resources is beginning to take its toll. The sharp decrease of most stocks in the north (an estimated 92% of commercial species in the Mediterranean are overexploited) is causing ships to head for the south.

Fish such as the Patagonia Toothfish and the Mackerel Icefish are of course targeted. The latter's specific quota allocated for the South Georgia area was not reached (less than 1% of the quota in 2013-14 after some years in decline) suggesting that there is a depletion of that resource, a decline that could be partially linked to the accidental capture of alevins when fishing for krill [CAM 90]. Even the tiny krill are under pressure to support the needs of aquaculture and fish canneries in the north. After intense exploitation by Soviet ships between 1980 and 1990 (> 400,000 tons/year), catches went back down to under 100,000 tons/year again after the fall of the Soviet Union. New fishing techniques, that avoid damaging krill, led to a 40% increase in catches in 2010, reaching a reported 285,000 tons in the 2013-14 season. For the 2014-15 season, 611,000 tons were announced, risking eventually endangering the entire food chain (cetaceans, crabeater seals etc.), especially since the fishing zones were closing in on the coasts and wild fauna feeding areas. The Madrid Protocol, which expires in 2048, is not enough to efficiently protect all the fish and crustaceans. The future looks bleak, because many are calling for the opening of the Antarctic and its ocean to commercial activity and industrial "norms" and as such, the greed of all exploiters. Humanity needs the planet, the whole planet, whatever the consequences!

1.4. Dynamics of the discovery of Southern Ocean biodiversity

The southern fauna and flora, isolated from other oceans and continents, are very original. They were discovered little by little, with each expedition, and even today entire components of the Southern Ocean's biodiversity are poorly understood.

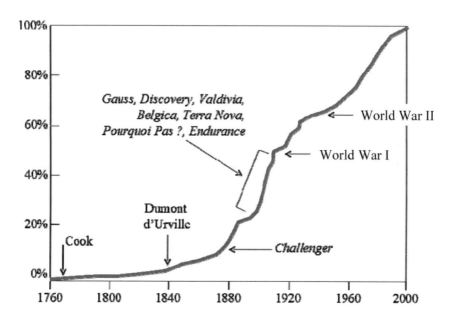

Figure 1.6. *Graph showing number of new species described in the Southern Ocean since 1760. Modified from [GRI 10]*

The first species described were found and brought back by 18th Century explorers, but they were relatively few in number. The knowledge curve initially progressed very slowly (see Figure 1.6). A few hundred species were identified in 1800, just under 400 when Dumont d'Urville landed in Adelie Land in 1840 and around 500 in 1859, the year that Darwin published *On the Origin of Species*. The pace picked up after the *Challenger* expedition (1873-76), but the greatest surge in the curve occurred during the boom preceding World War I. There was an increase from around 2,000 known species in 1890 to 4,400 by the eve of the war! Never before, nor after, has our knowledge of life in the Southern Ocean increased so fast. After two declines, in 1920 and 1940, progress became more regular and the number of described species was over 6,000 in the early 1950s and was around 9,000 in 2010.

The rhythm of discoveries has shown a slight decline these past few years, but the slope remains strong, suggesting that the inventory of Southern Ocean species is still far from over, even after the major international endeavor of the *Census of Antarctic Marine Life* (CAML), between 2005 and 2010. Rarefaction models predict, for example, that there are between 11,000 and 17,000 species of macrobenthos on the Antarctic continental shelf alone although only 5000 are known today [DEB 11]. Such projections are more realistic now that new genetic techniques have helped to uncover many cryptic species, even amongst supposedly well-known groups, and unsuspected symbiotic forms have also been found. Elsewhere, huge areas off the shores of Ellsworth Land and Marie Byrd Land remain to be explored. Finally, investigations of entire groups, such as nematode worms, sipunculid worms or echiurans are still just beginning [DEB 14].

1.5. Tools for oceanography exploration

Originally, the sole purpose of exploration was the mapping of the planet. Tracing the coastline and bathymetric studies of the seafloor contributed towards safer navigation. However, very soon the wish to create an inventory of species was added to the initial purpose. Sounding lines and spyglasses were supplemented by trawls and dredges to bring unknown specimens onboard research vessels (see Figure 1.3). With regard to physical oceanography, depth and water temperature measurements remained basic; the diversity of the sounding lines and thermometers used before 1890 bears witness to that fact. The speed of the currents remained unknown and was estimated through temperature measurements. Today, an oceanographer or marine biologist's arsenal has expanded considerably.

Ships are now equipped with multi-beam echo sounders that allow very precise mapping of the seabed, so that slopes can be calculated and the roughness of the seabed can be fully appreciated. In February 2013, the *Polarstern* thus discovered and mapped a new seamount in the middle of the Weddell Sea, at a depth of 400 m and rising to 25 m below the surface.

Niskin bottles, grouped in rosettes and equipped with sensors, allow measurements and samples to be taken at different depths (see Figure 1.7). It is possible to establish vertical profiles of temperature, salinity, dissolved oxygen, current speed and direction. Each sample of water brought up is analyzed.

Figure 1.7. *24 Niskin bottles in a rosette (approximate height of 1.50 m). At the top and bottom of each tube, the lids are open © Bruno David*

Near the Antarctic continent, measuring the amount of rare gases such as neon and helium allows the meltwater from the bottom of glaciers and ice sheets to be tracked on the very deep descent caused by its cold temperature. Measuring chlorofluorocarbons (CFCs), released by industry until they were banned, provides an indication of when the surface transit of the water analyzed occurred. Different water masses can thus be identified by their chemical signature and then mapped in time and space in order to understand their seasonal, or multiannual, evolution.

Various kinds of core drill are used to sample the sedimentary column at various thicknesses: a few dozen centimeters for studies of meiofauna (tiny fauna, between 1mm and 45μm), several meters for studies of sediment and a few dozen meters to reach relevant geological history of the past few million years. Other, more imposing methods, such as drillships that can produce cores from compact rock, are used to reach further back in geological history by drilling hundreds of meters into the ocean floor.

Planktonic flora and fauna are collected using nets with a very fine mesh. On the seabed, traps are used to catch larvae. For the smallest organisms (nano- or picoplankton, bacteria and viruses), water is collected in bottles and then filtered and analyzed using various techniques: flow cytometry, genetic markers, etc.

For biological samples, it all depends on depth. Within the first 30 to 50 meters, scuba diving is irreplaceable because it allows access to habitats in a manner almost as direct as on land. This method of investigation in the icy Antarctic waters flourished with the development of dry suits. Beyond these depths, trawls and dredges are often required. These have evolved relatively little over the past 140 years. Even though winches and deep-sea cables have undeniably progressed, the Agassiz trawl designed in 1880 remains just as effective. These traditional sampling tools have been joined by others (supra-benthic sleds, core drills, etc.) that have improved the quality of sampling, especially for smaller organisms. All of these instruments are now operated from oceanography research vessels with reinforced hulls, enabling them to penetrate through zones of sea ice that were previously inaccessible. These are complemented by exceptional tools that allow for ecological approaches, for the characterization of habitats or geology. The German Alfred Wegener Institute's Ocean Floor Observation System (OFOS) produces images and videos at the seabed to a resolution high enough for organisms to be identified and counted (see Figure 1.8), and is considered unrivalled by benthic ecologists. Manned submarines such as the French *Nautile* or the American *Alvin* transport the researcher to sites of interest, providing them with the opportunity to take samples of rocks, fluids or fauna, to take measurements and even to carry out experiments on site. Autonomous machines such as Remote Operated Vehicles (ROV) or Autonomous Underwater Vehicles (AUV) are currently in development because, despite costing far less than bathyscaphes, their remote controls are no less precise.

Figure 1.8. *Launching the OFOS seabed photography system from the Polarstern. Scale shown by human figure © Bruno David*

In contrast to these methods, which explore the depths, information transmitted by satellites tells us about the ocean's primary productivity, about its ice cover, about phytoplankton blooms, etc. Radiometric studies of the colors in the ocean provide indications about the levels of chlorophyll A, dissolved organic matter, particulate organic carbon, etc., which are valuable parameters for understanding and modelling the behavior of the ocean. The color of the ocean, whether observed in its natural state, or at varying wavelengths, or through different filters, has become an essential parameter for the discovery of its secrets.

2

The Southern Ocean and its Environment: a World of Extremes

The Antarctic continent is surrounded by the Southern Ocean, a vast stretch of marine waters whose characteristics, often extreme, are expressed as much in its geography and oceanography as in the environmental conditions prevailing there. The values for certain environmental parameters are at the outer limits of the wide range of climatic conditions present on the surface of the Earth: the temperature of seawater there, for example, is close to its freezing point. These extreme values have not, however, prevented life from adapting to such conditions and diversifying into a profusion of species. A broad range of oceanographic, geographical and environmental factors play a significant role in the distribution of marine species and communities. Knowledge of these factors is today essential for a better understanding of the distribution of species, their evolution and their adaptations [CLA 10]. Although extreme, the conditions prevailing in the Southern Ocean are much more varied than seemed to be the case at first. What was once called "the Antarctic ecosystem" has now been identified as several different marine ecosystems that biogeographers are striving to reference and map, while emphasizing how closely they are related [DEB 14]. The physicochemical parameters of these ecosystems have long been measured, but the development of new tools, such as satellite imaging, means that it is now possible to take continuous readings of certain parameters. These advances help to define the importance of the Southern Ocean in global ocean circulation. They will thus lead to a better understanding of the role of this ocean in the evolution of ancient climates, but also in modern climatic changes.

2.1. An ocean with undefined limits

Literally poles apart, the Arctic and Antarctic polar regions are situated in opposing geographical contexts. In the north, the Arctic Ocean occupies a central position. The geographical North Pole is in the middle of this ocean, where it is almost 4,000 meters deep, and the shores of the ocean correspond to the northern borders of the American and Eurasian continents. In Antarctica, it is the continent that occupies the polar position. The geographical South Pole peaks at an altitude of 2,835 m in the center of the continent, 1,300 km from the nearest coast. Antarctica is thus the most isolated continent on the planet. Its coasts are around 1,000 km from South America, 2,500 km from Australia and 4,000 km from Africa.

The Southern Ocean is a unique stretch of water circling the globe without interruption between latitudes 45° and 55° South, connecting the southern parts of the Atlantic, Indian and Pacific oceans. It can even be seen as the heart of the World Ocean, with the Atlantic, Indian and Pacific oceans as its northern extensions. The Southern Ocean does indeed play a central role in the circulation and renewal of global ocean waters [ORS 95]. In contrast with the Arctic, this ocean does not have clearly identifiable geomorphological limits and, since its hydrological borders are mobile (ocean fronts), no fixed, indisputable geographical boundaries can be defined that all countries will recognize. Since 2000, the Southern Ocean has been identified by the International Hydrographic Organization (IHO) as an ocean in its own right, which was not previously the case, bordered in the south by the Antarctic continent. To the north, however, the definition of its borders has not yet reached a consensus and so, depending on the country and the interests involved, the northern geographical limit of the Southern Ocean can either be a coastline (e.g. the south Australian coast, for Australians), or a parallel (between 60° S and 35° S, depending on the country). The 60° parallel south is thus recognized as the northern limit, but only by 14 countries (IHO circular letter of June 1st, 1999). Rather than adopting the official geographical limits motivated by each country's economic and political interests, oceanographers and biologists prefer the hydrological boundaries that represent natural borders, structuring marine biodiversity. For scientists, the Southern Ocean extends from the shores of Antarctica in the south to the Polar Front in the north. These were, in fact, the limits used by the Census of Antarctic Marine Life (CAML), an international program to inventory Antarctic marine biodiversity, and by the Commission for the Conservation of Antarctic Marine Living Resources (CCAMLR) [GRA 06]. When measured using the limits adopted by oceanographers and biologists, the Southern Ocean covers nearly 35 million km^2, or 10% of the planet's ocean surface [GUT 10]. In terms of area, it is the fourth largest ocean, after the Pacific, Atlantic and Indian oceans, but still ahead of the Arctic Ocean, which is by far the smallest (14 million km^2).

From a geomorphological point of view, the Southern Ocean is divided into three large ocean basins, with depths ranging from 4,000 m to 6,000 m: the Pacific, Indian and Atlantic basins. These basins are separated from one another by a series of ocean ridges: the Scotia, Mid-Atlantic, Southeast Indian, and Pacific-Antarctic ridges. There are also shallower submarine plateaus, the largest of which are Crozet, Kerguelen-Heard and Campbell (see Figure A.1 in the Appendix). At the edges of each continent, part of the Earth's continental crust extends beneath the ocean to form a continental shelf. Particularly extensive here, the continental shelf around the Antarctic covers 4.6 million km^2, or 11% of the total surface area of all the continental shelves on the planet [GUT 10]. With an average width of 125 kilometers, it is on average twice as wide as any other shelf [GRI 10]. It does, however, range in width from about 500 km in the Ross and Weddell seas to barely 15 km off Queen Maud Land. Another unique feature of the Antarctic continental shelf is its depth (450 m on average), two to four times greater than that of other continental shelves. This great depth is partly the result of intense glacial erosion that has cut deep canyons into the shelf and dug out internal basins (sometimes over 1,500 m in depth), but above all due to the continent subsiding beneath the weight of the ice that covers it. Indeed, the considerable mass of ice accumulated since the Last Glacial Maximum weighs down on the continent and has caused it to sink almost 700 m [CLA 10].

2.2. The southern climate: windy and cold, with very little light

2.2.1. *Strong winds*

In the Southern Hemisphere, the prevailing winds blow from subtropical high-pressure zones to the low-pressure belt surrounding Antarctica (see Figure 2.1). They are gradually deflected towards the east by the Coriolis effect (apparent drift due to the Earth's rotation) in such a way that they blow from west to east over most of the Southern Ocean. Where the winds reach their maximum intensity, they drive the Antarctic Circumpolar Current (ACC), the main ocean current responsible for the distribution and movement of water masses in the Southern Ocean. In the extreme south, the prevailing winds blow from east to west. They circle the Antarctic continent at about 65° S, generating the Antarctic Coastal Current, the second major marine current in the Southern Ocean [KNO 07]. Finally, very strong, offshore, katabatic winds can sometimes blow at gale force during the winter, from the continent to the sea, contributing to surface water motion and the formation of sea ice.

Figure 2.1. *Strong winds are common in Antarctica. South Shetlands* © *Bruno David*

2.2.2. *Extreme cold*

Low temperatures in the south can be explained by the low solar angle in polar regions, the high albedo of ice, which reflects light rays before they have time to heat the continent or the lower layers of the atmosphere, and the zonal atmospheric and oceanic circulation patterns, which thermally isolate the Antarctic from the rest of the planet, thus also keeping the temperatures low.

On the surface of the Southern Ocean, water temperature varies from 4 to 8°C in summer and from 1 to 3°C in winter in the Polar Front Zone. Closer to the continent, south of the Antarctic Divergence, surface waters remain close to freezing point all year round, i.e. from -1 to -1.9°C. There is, in fact, very little difference in temperature between surface water and deep water (less than 5°C). At the bottom of the ocean, the coldest waters are found in the East Antarctic and in the Ross and Weddell seas (0 to 2°C). These waters feed the Antarctic Bottom Water. In contrast, the Antarctic Peninsula region is under the influence of "warmer" waters (generally above 0°C), fed by circum-Antarctic deep-sea currents. This area also experiences the greatest seasonal variation. To the north, the temperature does not increase gradually, but changes rather abruptly perpendicular to the marine fronts separating the water masses. Sea surface temperature eventually rises to over 12°C to the north of the Sub-Tropical Front (see Figure 2.2).

2.2.3. *From winter night to weak daylight*

Like the Arctic, the Antarctic experiences strong seasonal variations in sunlight, with permanent night in winter and constant daylight in summer, south of the Antarctic polar circle (at latitude 66°33'S). Nevertheless, the average amount of daylight penetrating the water column and available for photosynthesis and the production of organic matter is insignificant, even in summer. This is explained by the low solar angle in the polar zone, by the predominance of cloudy weather conditions over much of the Southern Ocean, and by the presence of sea ice. To this must be added the classic conditions that affect all oceans, i.e. the reflection of light on the water surface (up to 50% of light-rays are reflected) and light absorption by suspended particles [KNO 07].

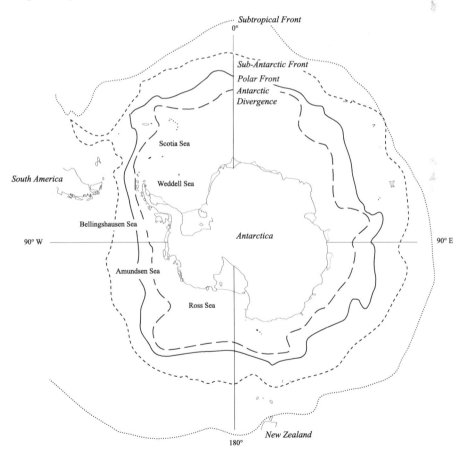

Figure 2.2. *Map of the Southern Ocean representing the main oceanic fronts. Modified from [DEB 14]*

2.3. Ice in all its forms

2.3.1. *Sea ice*

At the surface of polar seas, water freezes when it reaches a temperature of -1.9°C, the freezing point for a salt content of 35 grams per liter, and not at 0°C, as is the case for freshwater. When the sea freezes, sea ice is formed. In contrast with the Arctic Ocean, where sea ice remains present throughout the year (although the amount of ice has decreased considerably over the past decade), the ice formed in winter on the Southern Ocean melts almost completely in summer, a phenomenon unrelated to global warming. Thus, as the austral summer ends (at the end of February), only 3 million km^2 of ocean remains frozen, mainly in the east of the Weddell Sea and in the Bellingshausen and Amundsen seas. In comparison, as the austral winter ends (at the end of September), the ice extends over nearly 18 million km^2 (almost twice the size of Europe). This major seasonal contrast is one of the largest physical changes to occur each year on the surface of our planet.

Sea ice is generally divided into two types: drift or pack ice, and land-fast or shore-fast ice. Pack ice is seasonal, drifting at the ocean surface, pushed by marine currents and strong winds all through the austral winter. It usually measures less than a meter, rarely more than two meters, in thickness. Fast ice is formed by sea ice that clings to the coastline, on shoals and grounded icebergs. It remains contiguous with the coast and can persist for several years, forming multi-year strata that can reach a total thickness of several meters (see Figure 2.3). It is the main physical barrier for Antarctic coastal marine ecosystems, determining the development of benthic communities in the depths (see Chapter 4).

During the austral winter, some areas between the drift ice and the continent never freeze. These are the polynyas, found mainly in the coastal areas of East Antarctica. These open water areas form in places where strong continental winds blow (katabatic winds) and, together with marine currents, carry the ice out to sea as it forms. Polynyas play an important role in heat transfer between ocean and atmosphere and are veritable ice factories. They also participate in the formation of the cold dense waters characteristic of the Antarctic continent. Although seawater begins to freeze at -1.9°C, it is a slow process, as saltwater becomes denser as it cools, thus sinking deeper down, and avoiding transformation into ice. The water replacing it at the surface is in turn subjected to the cold atmosphere and starts to freeze. Furthermore, as ice forms, it cannot retain all the salt originally present in seawater. At best, sea ice retains only 22 g out of the 35 g of salt present per liter of seawater. This excess salt is carried away as brine flowing down through the sea ice, so that the ice at the surface tends to desalinate over the winter, and may even be drunk by explorers who have lost their way! Meanwhile, the excess salt at the base of the sea ice increases the salinity of the underlying cold waters, thus causing them to become even denser. Brine rejection contributes to the formation of zones of cold, dense saltwater on the Antarctic continental shelf, waters that will then supply the Antarctic Bottom Water. Finally, polynyas also have an important ecological role: their existence favors organic matter production by phytoplanktonic microalgae and participates in the functioning of coastal ecosystems.

Figure 2.3. *Ice shelf ending in a cliff at the ocean's edge. Below, the sea is covered in multi-year fast ice. Large grounded tabular icebergs are visible in the background. Atka Bay, Weddell Sea © Thomas Saucède*

2.3.2. *Ice sheets and ice shelves*

The continent of Antarctica is almost entirely covered by a thick ice sheet, with an average thickness of 2,200 m. In some places, the ice sheet can exceed 4,000 m in thickness, the record being 4,806 m, measured between the Russian Vostock station and the French-Italian Dôme C station in East Antarctica [REM 03]. This immense volume, 30 million km^3, represents 90% of the ice present on the planet today, or 75% of the world's freshwater reserves. Along almost half of the Antarctic coastline, the ice sheet is detached from the bedrock and thus floats on the ocean, creating vast ice shelves. As bedrock frequently lies below sea level, buoyancy eventually overcomes gravity and the ice sheet begins to float several meters to several hundred meters above the seabed, especially since the ice sheet is often very thin near the coast (maximum 300 m thick). These ice shelves alone represent nearly 10% of the total surface area of the Antarctic ice sheet. Two ice shelves are particularly extensive: the Ross Ice Shelf has a surface area of 526,000 km^2 and the Ronne-Filchner Ice Shelf in the Weddell Sea measures over 473,000 km^2 [REM 03]. Each of these shelves is therefore comparable in area to metropolitan France!

Such ice shelves create incredible landscapes at the ocean's edge. Steep cliffs rise to over 30 meters above sea level, which still leaves 270 meters of ice underwater (see Figure 2.3). These ice cliffs regularly break off into the sea as the ice shelf moves away from the continent and becomes exposed to wave action. This phenomenon, known as iceberg calving, produces characteristic tabular forms that can reach colossal dimensions, so large that explorers have sometimes confused them with actual islands. The largest tabular iceberg ever seen was probably Iceberg B-15, which broke off from the Ross Ice Shelf in 2000. It measured 295 km long and 37 km wide, i.e. almost 11,000 km^2, which is much larger than Corsica! The lifespan of a tabular iceberg is about four years, on average, but a huge iceberg (measuring 66 km long and 18 km wide), again from the Ross Ice Shelf, which broke off in 1992, was followed by satellite for eight years as it drifted all the way to the Argentine coast. Under the effect of increasing air temperatures, some ice shelves have recently collapsed, like the Larsen Ice Shelf in West Antarctica, in 1995, and again in 2002. Its collapse produced many icebergs that have had a profound impact on marine communities and ecosystems [GUT 01, GUT 11]. Indeed, as they drift, large icebergs can run aground on shoals, scraping the sea bed, and ravaging the communities living there.

2.4. In isolation yet interconnected, the complexity of ocean circulation

2.4.1. *Ocean currents*

2.4.1.1. *The Antarctic Circumpolar Current*

From an oceanographic point of view, the Southern Ocean can be defined as the region of the World Ocean that is under the direct influence of the Antarctic Circumpolar Current (ACC). It is the only current to flow all the way round the planet, completely encircling Antarctica, since no land impedes its journey around the world between 45° S and 55° S. The ACC measures 24,000 km in length and ranges in width from 200 km (south of Australia) to 1,000 km (south of the Atlantic). It narrows along the Drake Passage, between the Antarctic Peninsula and South America, and its course is diverted by the underwater landscape (ridges and submarine plateaus). If we were to follow the journey of a single drop of water within this current, we would find that on average, it takes three and a half years to travel the entire length of the ACC, but that drop would have to circumnavigate Antarctica six times before it could escape from this very homogenous flow, and would therefore have been held prisoner for 21 years! With a flow rate of between 134 and 164 Sverdrups (Sv) at the Drake Passage (one Sverdrup corresponds to the flow rate of one million cubic meters per second, or the cumulative flow rate of all the rivers in the world), the ACC is by far the largest current on Earth. In comparison, the Gulf Stream transports only 30 Sv along the coasts of Florida, which is still 150 times the flow rate of the Amazon! Given the colossal volume of water that it transports, the ACC inevitably plays an important role in both global ocean circulation and the planet's climate. Although at the surface it isolates Antarctic waters and the Antarctic continent from the main sub-tropical gyres, at depth it drives the vertical flow of water masses, taking part in the ventilation and renewal of abyssal waters. Thus, in the Polar Front Zone, it contributes to downwelling, the sinking of surface water rich in oxygen and carbon dioxide, while in the Antarctic Divergence Zone, it drives upwelling, the rise to the surface of nutrient-rich (phosphates, silicates and nitrates) deep sea waters. Through these movements, the ACC contributes to the global thermohaline circulation regulating the world's climate.

2.4.1.2. Antarctic Coastal Current

At around 65° S, winds create a divergent zone between water masses called the Antarctic Divergence. South of this latitude, surface waters no longer flow east but westward, forming the Antarctic Coastal Current. This current flows along the Antarctic coasts, making two deep incursions into the Ross and Weddell Seas, where it joins the southern part of the spiral currents (gyres) present in these seas. Unlike the ACC, the Antarctic Coastal Current does not completely encircle the continent but is interrupted in West Antarctica, along the Antarctic Peninsula and the Drake Passage (see Figure 2.4). Furthermore, its flow rate is much lower (10 Sverdrups on average). It covers an area of high surface water primary productivity (photosynthesis by phytoplankton) in the Ross and Weddell Seas as well as on the continental shelf, facilitating the spread of species along the coast.

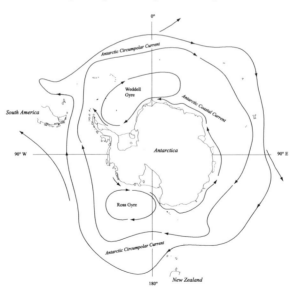

Figure 2.4. *Map of the Southern Ocean, showing the main surface currents. Modified from [DEB 14]*

2.4.2. Ocean fronts and the zonation of water masses

In the Southern Ocean, the importance of the Antarctic Circumpolar Current is demonstrated by the latitudinal separation of water masses by their physical, chemical and biological properties. The physicochemical transitions between water masses are abrupt and marked by ocean fronts that limit mixing between northern and southern sectors. Ocean fronts are narrow, relatively stationary zones, where steep gradients of variation can be observed in oceanographic parameters (e.g. temperature, salinity, etc.). The structure of the Southern Ocean is complex, and as many as nine ocean fronts have been identified but, as water circulation is sometimes disturbed (e.g. at the edges of the Kerguelen Plateau), it is difficult to identify these fronts in some places [ORS 95]. Basically, four major oceanographic boundaries structure the ACC and the Southern Ocean (see Figure 2.2): the northern limit is determined by the *Sub-Tropical Front*; its central part is structured north to south by the *Sub-Antarctic Front* then by the *Antarctic Polar Front*; finally, its southern limit is defined by the *Antarctic Divergence*. In the depths, oceanographic gradients are less pronounced, but there is vertical zonation of the water masses in addition to latitudinal zonation (see Figure 2.5).

At the bottom of the Southern Ocean, the waters are generally very cold, with temperatures below 1°C, with few exceptions (such as the bottom waters of the Kerguelen Plateau). The main contrasts are seen around the continent between the Antarctic Peninsula, which has relatively "warm" bottom waters, reaching temperatures of 1 to 2°C, and the East Antarctic and the Weddell Sea, where water temperatures are always below 0°C. The bottom waters of the Peninsula are in fact influenced by the upwelling of warmer circum-Antarctic deep waters, while the Weddell Sea and the East Antarctic are fed by deep, particularly cold, Antarctic waters.

2.4.3. A complex interplay between wind, water and ice

The main exchanges between water, air and ice occur in the surface layers of the Southern Ocean, where the waters acquire their essential physicochemical characteristics (i.e. temperature, salinity, oxygen, etc.). Unlike surface marine currents mainly driven by strong winds (westerly winds for the ACC, easterly winds for the Antarctic Coastal Current), vertical movements of water masses depend on their relative densities, controlled by their temperature and salinity. The saltier and colder the water masses, the denser and more likely they are to sink under warmer, less dense and less salty water. Ice is also involved in ocean circulation. All year round, through contact with ice shelves, Antarctic surface waters become colder, denser and sink northward along the continental shelf, then down the Antarctic continental slope to supply the Antarctic Bottom Water that then flows into other basins further north. A similar phenomenon occurs in winter near the sea ice. Antarctic Surface Water cools as it encounters ice and polar air, and becomes more saline due to brine discharge as sea ice forms. This cold, hypersaline surface water becomes denser and sinks to replace the bottom water of the Antarctic continental shelf.

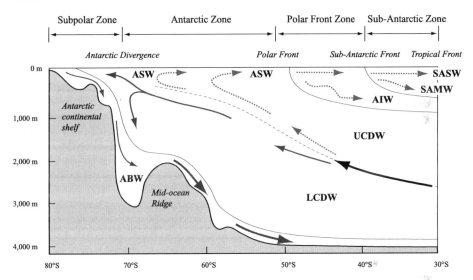

Figure 2.5. *Vertical and latitudinal zonation of water masses in the Southern Ocean. Modified from [DEB 14]. The dotted grey arrows represent the movement of the warmest waters, the black arrows show that of the coldest waters. The Sub-Tropical Front marks the northern boundary of the Sub-Antarctic Zone (beyond that limit, surface water temperature is above 10-12°C) and the northern limit of the area affected by the ACC. The Sub-Antarctic Front (surface water temperature above 5-7°C, further north of this limit) separates Sub-Antarctic Surface Water (SASW) and Sub-Antarctic Mode Water (SAMW) from Antarctic Intermediate Water (AIW). The Antarctic Polar Front or Antarctic Convergence (to the south, surface water temperature below 4°C) separates Antarctic Intermediate Water from Antarctic Surface Water (ASW) which extends south to the coast. These water masses overlie the Upper and Lower Circumpolar Deep Water (CDW). The area between the Polar Front and the Sub-Antarctic Front is the central part of the ACC. South of the Polar Front is the Antarctic Zone. The front to the south of the ACC or Antarctic Divergence (surface temperatures between -1.5°C and 0°C) corresponds to the limit between the waters of the Antarctic Zone to the north and the Sub-Polar Zone (or Antarctic Coastal Zone) to the south. The Sub-Polar Zone is the region affected by the Antarctic Coastal Current, and it is where Antarctic Bottom Water (ABW) is produced*

2.4.4. *Water in action*

The strong westerly winds that blow continuously at the surface of the Southern Ocean force the surface water masses to flow eastward, creating the ACC. For physical reasons, ACC flow generates a convergent movement of the Sub-Antarctic and Sub-Tropical Surface Waters towards the Sub-Tropical Front in the north and, conversely, a divergence of Antarctic Surface Water on the southern border. The relative void created by the divergence is offset by the rise of Circumpolar Deep Water in a phenomenon known as upwelling.

Two thirds of Circumpolar Deep Water rising at the Antarctic Divergence flows north and feeds into Antarctic Surface Water (see Figure 2.5). This water is cold, low in salt content and rich in dissolved oxygen and nutrients. It flows all the way to the Polar Front, where it sinks to join the Antarctic Intermediate Water. A third of Circumpolar Deep Water flows south to the edge of the Antarctic continent. There, it cools down, sometimes increasing in salt content, becomes denser and sinks to the bottom of the ocean. This very cold and salty bottom water then circulates north (mainly from the Weddell Sea) to feed into the Antarctic Bottom Water, and then into the deep water masses of other oceans.

The complex interplay between wind, water and ice thus generates movements with contrasting effects between surface and deep water masses. By circulating westward around the Antarctic continent south of the Antarctic Divergence, but eastward to the north of this area, Antarctic Surface Water tends to isolate the continent and its coastal waters from the effects of other oceans by limiting north–south exchanges. It also plays a role in homogenizing the Antarctic Surface Water's hydrological characteristics around the continent. In the depths, however, the movement of water masses has the opposite effect. By flowing northward along the continental shelf and slope, the Antarctic Bottom Water connects the different ocean basins with one another and participates in the thermohaline circulation of the World Ocean that governs Earth's climate through heat transfer between ocean and atmosphere.

2.5. Sediment and nutrients

2.5.1. *Marine sediment and its origins*

The nature of the sediment covering the ocean floor (soft, fine or coarse sediment) is a determining factor in the settlement of communities and the functioning of marine ecosystems [DEB 14]. Marine sediment is either terrigenous (derived from the erosion and chemical weathering of continental rocks) or biogenic (formed out of a multitude of calcareous or siliceous skeletal remains of planktonic or benthic organisms).

In the Southern Ocean, terrigenous sediment is mainly the product of glacial erosion and a large proportion of the sediment found on the seabed provides evidence of the highly erosive action of coastal glaciers during the Last Glacial Maximum, around 20,000 years ago. This sediment has accumulated at the mouths of ancient glacier valleys that sometimes cut deep into the Antarctic continental shelf. More generally, fine sediment settles in hollows on the continental shelf, at the base of the slope and on the abyssal plains, while coarse sediment is found at the top of the slope and on the shoals of the continental shelf, deposited by strong bottom currents and by icebergs.

The majority of biogenic Antarctic sediment is siliceous in nature. Antarctic surface waters in the sub-polar region are in fact very rich in diatoms, microscopic organic plankton with siliceous skeletons (see Figure 2.6). Diatoms make their skeletons from dissolved silica available through the upwelling of Antarctic Deep Water, rich in nutrients, particularly silica. When diatoms die, their siliceous skeletons become part of the ocean floor as particulate silica, forming substantial sedimentary silica deposits (opal) between 50 and 70° S. Moreover, a significant portion of dissolved silica in deep waters is used directly by sponges, which form a living reservoir of silica on the Antarctic continental shelf. The Southern Ocean is a therefore a source of silica for the World Ocean, and 75% of all siliceous sediment in the ocean has accumulated between the Polar Front and the Antarctic continental shelf [TRE 14].

2.5.2. *Oxygen and nutrients, sources of marine life*

The production of organic matter by microalgae in phytoplankton is the basis of food chain webs, and thus of energy flow and the functioning of Antarctic marine ecosystems. It is the result of photosynthesis, which is only possible in the presence of light and nutrients. Organic matter produced by phytoplankton is in part consumed in the water column by various organisms, from the consumers of microalgae, such as krill and zooplankton (copepods and salps), to the final consumers, i.e. large carnivores such as orcas and seals. Only 3–10% of the organic matter produced in surface water reaches the ocean floor. At the bottom, part of the organic matter is ingested by benthic organisms, while the rest is re-mineralized and returned as macro-nutrients to the water column through upwelling.

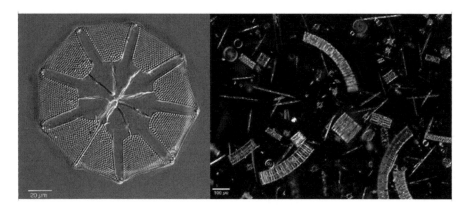

Figure 2.6. *Diatoms collected in the Amundsen Sea. Diatoms are photosynthetic unicellular algae with siliceous skeletons. They are part of the phytoplankton. Their skeletons form siliceous sedimentary deposits found between latitudes 50 and 70° S © CNRS Photo library/John Dolan*

In the ocean, the oxygen content of surface waters is high due to exchanges between ocean and atmosphere. In contrast, water that has long been in the depths is depleted of oxygen. Global ocean circulation supplies these deeper zones with oxygen-rich water when surface waters sink. Conversely, in upwelling zones, nutrient-rich bottom waters rise to fertilize the surface and increase the production of organic matter by phytoplankton. In the Southern Ocean, the Antarctic Divergence is where upwelling mainly occurs, with water from other oceans rich in macro-nutrients, nitrates, phosphates and silicates (Upper and Lower CDW). Most of this water then flows north, so that the concentration of macro-nutrients in Antarctic Surface Water remains high until it reaches the Polar Front, where it sinks to become Antarctic Intermediate Water (see Figure 2.5). Beyond the Sub-Antarctic Front, surface waters are depleted of macro-nutrients.

Although Antarctic Surface Water is rich in oxygen and macro-nutrients, primary production of organic matter by phytoplankton is limited [GRI 10]. This apparent paradox is explained by the lack of light energy and the small amount of dissolved iron available, which are two key factors in photosynthesis and primary production. Iron concentration in surface waters is highest north of the Polar Front, in the Sub-Antarctic Zone, and south of the Antarctic Divergence, near the continent. This region is the only one to unite all the conditions necessary for the development of phytoplankton, as its waters are rich in macro-nutrients, iron and oxygen. As a result, the coastline and the Antarctic continental shelf are the areas with the highest organic matter production. This general pattern can be modified to some extent as, under the ACC effect, the presence of upwelling zones downstream of large underwater escarpments generates waters locally enriched in dissolved iron. This explains why areas to the east of the Drake Passage, the Crozet and Kerguelen plateaus and the submarine ridges are also high in primary production [DEB 14].

3

The Ocean Through Time

The distribution of species in the Southern Ocean is not determined solely by present environmental and oceanographic conditions. The originality and richness of Antarctic marine biodiversity also reflect a long evolutionary history, closely linked to the geological and climatic evolution of the Southern Ocean [CLA 89]. In order to understand the distribution and evolution of existing marine species in the Southern Ocean, it is not enough to know the ocean of today: it is also necessary to understand its evolution.

Antarctica has occupied its current latitudinal position, deep in the Southern Hemisphere, for millions of years. The continent has moved very little since the middle of the Paleozoic Era, 390 million years ago. The main elements in its tectonic history that have led to its present geography consist of the successive stages in the break-up of the supercontinent, Gondwana, which united Antarctica, Africa, South America, India, Madagascar, Australia and New Zealand during the Jurassic Period [KNO 07]. During the last 180 million years, the evolution of the Southern Ocean has brought about three-fold isolation: the geographic isolation of the continent, following the tectonic fragmentation of Gondwana, its oceanographic isolation, with the establishment of the Antarctic Circumpolar Current and its oscillations, and climatic isolation, leading to the glaciation of the continent, which began well before the first ice appeared in the Northern Hemisphere.

The tectonic, climatic and oceanographic events that shaped the Southern Ocean are closely linked, because the processes governing these changes are complex and interconnected, with many feedback loops [RUD 02]. Two tectonic events have played a crucial role in the recent history of Antarctica, by opening two major marine passages: the Drake Passage, between South America and Antarctica, and the Tasman Gateway, between Antarctica and Australia. The opening of these two passages allowed the establishment of deep ocean water circulation, considered to be the source of subsequent changes in global ocean circulation and the climatic evolution of the planet. They were, in part, the cause of Antarctic glaciation and the isolation of flora and fauna that would evolve separately from the rest of the Earth's biosphere.

3.1. The split of a supercontinent from the Jurassic to the Eocene

Until the end of the Early Jurassic, 180 million years ago (see Figure A.2 in the Appendix for the Geologic Time scale), Antarctica was attached to the other continents in the Southern Hemisphere, forming a single continental landmass, Gondwana (see Figure 3.1). Gondwana began to separate into an eastern block and a western block, at the very end of the Early Jurassic, around 175 million years ago [ALI 08]. A narrow ocean channel formed between Madagascar and Africa, and then extended gradually south during the Early Cretaceous (120 Ma), between the western part (Africa and South America) and the eastern part (Antarctica, New Zealand, Australia, India and Madagascar) causing them to separate. On the eastern side, India and Madagascar both moved away from Antarctica and Australia, at the end of the Early Cretaceous (100 Ma). However, the similarity of the fossil species (dinosaurs, in particular) found on these different continental blocks suggests that some land routes must have remained, and so species dispersal may have occurred via the partially emerged Kerguelen plateau as it formed [ALI 08]. Around 80 million years ago, in the middle of the Late Cretaceous, Antarctica, Australia and New Zealand had completely separated from the other continents. New Zealand, made up of several different parts, moved away from Antarctica at the end of the Cretaceous, between 80 and 65 million years ago.

In the west, at the end of the Cretaceous Period, tectonic movements in Africa, South America and Antarctica resulted in the convergence of the Antarctic Peninsula and Patagonia. An ephemeral (at the geological timescale) land bridge, known as the Weddellian Isthmus, formed between the two continents. This isthmus lasted around 15 million years, from the end of the Campanian (70 Ma) to the Paleocene (55 Ma) [REG 15]. The presence of this isthmus had considerable impact on the biodiversity of the time by allowing exchanges between the two continents. This was the case for the first large mammal fauna, such as the ungulates, Litopterna, a well-diversified order of hoofed herbivores, now extinct. From the end of the Paleocene, the Weddellian Isthmus was flooded and replaced by a shallow channel [REG 15], but the Antarctic Peninsula and Patagonia did not completely move apart from each other until the end of the middle Eocene, around 40 million years ago [CUN 95]. From then onwards, the waters deepened and the Drake Passage formed, as shown by the differences between Patagonian and West Antarctic marine fossils [DEB 14].

Figure 3.1. *Paleo-geographic map of the Southern Hemisphere in the Early Jurassic (180 Ma) centered on the South Pole. The continents were united in a supercontinent, Gondwana. Landmasses are shown in gray; continental shelf limits are indicated by a solid black line. Redrawn from [LAW 03]*

On the other side of Antarctica, at the beginning of the Eocene (50 Ma), Australia began to separate from Antarctica and the Tasman Gateway opened. Initially, the channel was relatively shallow, as the Australian and Antarctic continental shelves were still joined. The shallowness of the channel prevented any exchange between Pacific and Atlantic waters, preventing the development of a circumpolar current [LIV 05]. The Tasman Gateway deepened significantly at the end of the middle Eocene, 40 million years ago, as Tasmania and Australia moved northeastward away from the Antarctic (see Figure 3.2). From then onward, cold surface currents were able to flow around the Antarctic continent [EXO 04]. With the opening of the Drake Passage in the west and the Tasman Gateway in the east, circumpolar circulation became possible and a proto-Antarctic Circumpolar Current (ACC) developed. Antarctica and its biodiversity became geographically isolated from other continents. The evolution of Antarctic species thus occurred mostly in isolation from the fauna and flora of lower latitudes, giving rise to some part of their originality. Oceanographically, Antarctic coastal waters were no longer influenced by the warmer ocean waters to the north [LAG 09].

The development of isotopic methods of analysis now allows paleoclimatologists and paleo-oceanographers to reconstruct with precision the evolution of ocean water temperature and circulation patterns over the course of geological time. These analyses are based on samples of marine sediment and organic remains (e.g. foraminiferal tests or fish teeth) that provide natural climatic and oceanographic archives for scientists [DEC 14].

From the end of the Paleozoic to the Eocene, our planet experienced a period of nearly 190 million years when hot climates prevailed globally. These climatic conditions are known as *greenhouse* periods, in reference to the "greenhouse effect". Such hot climates can be explained by the high levels of greenhouse gases in the atmosphere at the time, resulting from intense volcanic activity at ocean ridges. Although greater quantities of CO_2 were emitted into the atmosphere, only a small part was used or "drawn down" by the chemical weathering of rocks, as there was little elevated terrain on the continents. This hot period was also characterized by a high overall sea level and by considerable expansion of epicontinental seas (also known as shelf seas). These vast marine areas absorbed more sunlight and thus enabled the planet to store more heat. The latitudinal temperature gradient was less marked than it is today, with high tropical temperatures decreasing very gradually towards the poles. Tropical or subtropical influence extended to high latitudes (50° S at least), not only at the surface of the ocean but also in the depths, whilst polar regions benefitted from a mild climate, with heat transported by warm subtropical waters. Antarctic surface waters were, however, cooler and denser than those further north, and sank into the depths to drive deep ocean circulation. Deep ocean circulation nevertheless had far less impact on global climate at that time than it does today [RUD 02]. Throughout the entire greenhouse period, the southern margins of Gondwana experienced a temperate climate and there was no ice cap [ZIN 82].

After reaching the Cretaceous Thermal Maximum during the Cenomanian, a period also corresponding to the highest global sea level ever recorded in over 250 million years, sea temperatures began to fall and global ocean circulation was modified. Ocean circulation at that time was very different from today, as it was latitudinal, and dominated by east-west flows. The opening of the Atlantic Ocean caused a gradual shift to meridional circulation (north-south flows). Many deep-sea drilling programs (notably within the Ocean Drilling Program) carried out in all oceans have studied the evolution of ocean temperature over geological time. Results reveal a dual phenomenon: a long period of cooling from the Cenomanian to today, interrupted by some irregularities, with warmer phases of variable length (see Figure 3.3). This general cooling trend is mainly explained by the uplift of mountain ranges during the Alpine Orogeny (i.e. the Alps, the Himalayas and the Andes), while chemical weathering of these young mountains increased the "draw-down" of atmospheric CO_2.

Figure 3.2. *Paleogeographic map of the Southern Hemisphere in the Eocene (40 Ma) centered on the South Pole. Landmasses in gray, continental shelves and ocean plateaus, black outlines. Redrawn from [LAW 03]*

The irregular nature of this cooling phenomenon first appeared in the mid-Paleogene Period, 55 million years ago, when the Antarctic climate experienced a sudden strong increase in temperature, with ocean surface waters heating by 4 to 6°C in less than 10,000 years. It was a brief episode, but of global magnitude. It lasted around 200,000 years, during the Paleocene-Eocene transition, and corresponds to the Paleocene Eocene Thermal Maximum. Subtropical waters at that time extended far into the south, probably reaching the Antarctic coast [STO 90].

The climate remained warm during the early Eocene [DEC 14], but then the temperature of deep ocean water began to decrease steadily. At the beginning of the

Eocene, the temperature of ocean water masses was relatively homogeneous, with an average of around 15°C. At the end of the Eocene, this had decreased to 11°C at the surface and 9°C in the depths. At the end of the middle Eocene, the transition began from a *greenhouse*-type climate to a colder, *icehouse* climate. In Antarctica, temperatures fell sharply, especially in the eastern part, thus leading to the initial glaciation of the continent [ARO 09].

3.2. Global cooling at the Eocene-Oligocene transition

The fossilized remains of organisms preserved in marine sediment (foraminifera, radiolaria, and diatoms) provide evidence of a sharp drop in global ocean temperatures during the Eocene-Oligocene transition, around 34 million years ago. It is estimated that the temperature of Antarctic waters fell by 6°C, with surface water temperature close to 0°C near the Antarctic coasts [ZAC 08, DEC 14]. It was also at that time that the Antarctic Bottom Water was generated, and deep ocean circulation began. Many authors have suggested that the opening of Drake Passage to deep circumpolar circulation reinforced the proto-ACC, forming the Polar Front, and thus leading to the thermal isolation of Antarctica. The continent cooled down because it no longer benefitted from the transfer of heat from lower latitudes [LAG 09]. However, the proven existence of shoals has led to partial reconsideration of this viewpoint. The ancient volcanic arc of the South Sandwich Islands (between Patagonia and the Antarctic Peninsula), and the Kerguelen and Broken Ridge plateaus in the Indian Ocean (today separated by about 2,000 km), must have formed shoals blocking the flow of deep circumpolar circulation, and thus limiting the strength of the ACC and its climatic consequences at that time [DAL 13b]. Other factors must therefore have intervened and contributed to the cooling of the World Ocean, particularly in the Antarctic. The effects of our planet's orbital parameters and a decrease in the atmospheric concentration of CO_2 are the most likely factors [LEF 12]. Reduced obliquity (tilt of the Earth's rotational axis in relation to its orbital plane) would have resulted in colder summers at higher latitudes. In addition, decreased greenhouse gas concentration in the atmosphere, resulting from an increase in the draw-down of atmospheric CO_2 through the chemical weathering of Alpine and Himalayan mountains, could also have contributed to the cooling. The intensification of the ACC and the thermal isolation of Antarctica would therefore be the consequence of these atmospheric and climatic changes, and not their cause [LEF 12]. Other hypotheses have also been put forward, but they remain tentative and are still under debate. This is the case with the asteroid impacts that occurred at the same time as the cooling of the planet. These impacts created the Popigai crater in Siberia (100 km in diameter) and the Chesapeake Bay crater in the United States (90 km in diameter) [DEC 14].

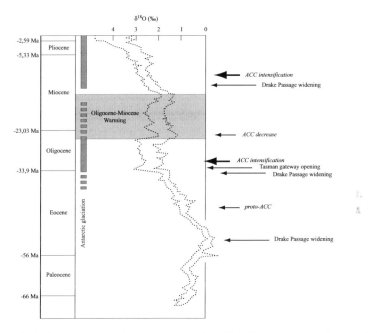

Figure 3.3. *Evolution of oxygen isotope values ($\delta 18O$ in ‰) during the Cenozoic Era (the two curves bound the confidence interval). The more the values increase, the more seawater temperature decreases. The general cooling trend is punctuated by irregularities and warming phases. The extent of Antarctic glaciation and the main tectonic and oceanographic events are plotted along the curve. Redrawn from [ZAC 01] and [LAG 09]*

The major cooling event at the Eocene-Oligocene boundary coincides with the oldest indications of glaciation in the Cenozoic Era. Such indications include glaciomarine sediment called IRD (Ice-Rafted Debris) formed from the scattered debris of continental rocks wrapped in fine marine sediment. Only one process could lead to such deposits: the erosion of continental rocks by glaciers, followed by the transport of the debris into the ocean, attached to icebergs, before its release on to the seafloor as the icebergs melt. Such IRD is therefore the first evidence of Antarctic glaciation, nearly 30 million years before it began in the Northern Hemisphere (the glaciation of Greenland is dated somewhere between 3 and 7 million years ago). Finding IRD thus implies that icebergs were drifting on the ocean and that glaciers were present on the Antarctic continent. Some authors even believe that a major ice sheet may have developed as early as this [SIE 09] and that the ocean must have been partly covered by sea ice in the winter [DEB 14]. Through a feedback loop, this initial glaciation would have contributed to the global cooling of the planet by increasing planetary albedo, since the ice would reflect the Sun's rays [ZAC 01].

3.3. Other thermal anomalies during the Oligocene and Miocene

The cooling that began at the end of the Eocene lasted throughout the Oligocene. Small ice sheets then spread over the entire surface of the Antarctic continent. At the end of the Oligocene, a new phase of global warming occurred. Unsurprisingly, this event was linked with the tectonic narrowing of the Drake Passage, which led to a weakening of the ACC and therefore a reduction in the thermal isolation of the Antarctic continent and its ocean [LAG 09]. Warmer temperatures became stable and then rose again during the Mid-Miocene Climatic Optimum. This episode corresponds to a glacial minimum, dated around 20 million years ago, during which an ice sheet could perhaps have persisted to the east in the center of the continent, while the shores benefitted from a temperate climate (without iceberg formation). In West Antarctica, only the Ellsworth mountain range (where Mount Vinson, the tallest summit in Antarctica, at 4,892 m in altitude, is located) and the Whitmore mountains would have retained their glaciers [DEB 14].

3.4. Another cold snap in the late Miocene

The end of the Miocene saw a new episode of cooling on a global scale. Deep ocean water temperatures fell by several degrees. In the Southern Ocean, this episode was marked by peak productivity of siliceous plankton in surface waters [KNO 07]. The ice sheet developed significantly over the entire Antarctic continent. The ice sheet was more extensive in East Antarctica, and scientists have estimated that it might have extended further than it does today [PAS 11]. A thick ice sheet also formed in the west. It was at this time that the Andean glaciers began to develop in Patagonia and the glaciation of Greenland first began. These heavy accumulations of ice had global repercussions. Increased storage of water as ice led to a considerable decrease in average sea level. Many submarine canyons were formed, which today still cut deeply into continental shelves around the world.

This global cooling can be explained by a new phase of intensification for the ACC, as the result of much greater opening of the Drake Passage and the Tasman Gateway. This intensification of the ACC can also be explained by tectonic movements in other parts of the world. The closing of tropical marine passages between New Guinea and Australia, before the end of the middle Miocene, and between the two Americas (Isthmus of Panama) in the Pliocene, caused the deflection of strong marine currents that served to reinforce the ACC in the south [LAW 03]. The cooling event was also associated with a decrease in the concentration of atmospheric CO_2. The cause, in this case, was not tectonic (Alpine ranges), but oceanographic. The establishment of deep ocean upwelling in the East Pacific would have stimulated production of organic matter by plankton, thus removing carbon from the atmosphere (and carbon is a component of CO_2). This organic matter did not degrade (the carbon remained trapped) and was preserved in sedimentary deposits, such as those of the Monterey Formation in California [DEC 14].

3.5. Climatic oscillations and glacial-interglacial cycles

From 5.3 to 2.6 million years ago, the Pliocene underwent a phase of global warming. Then, from the end of the Pliocene, global climatic evolution was characterized by a succession of fluctuations controlled by the Earth's astronomical cycles (obliquity, precession and eccentricity). In the long term, these fluctuations were part of the cooling trend that included two particularly cold events, at 2.5 million years ago and 700,000 years ago. In Antarctica, these events were marked by the alternating advance and retreat of ice sheets on the continental shelf as well as by cycles of sea-level variation. The ANDRILL drilling program, coring the rocks under the Ross Sea ice shelf, revealed that there have been no fewer than thirty-eight successive glaciomarine cycles over the last five million years, amongst which there were eleven major glaciation events [NAI 09]. During periods of maximum glaciation, the ice shelves rested on the seabed and extended to the edge of the Antarctic continental shelf (see Figure 3.4). Down below, the slope linking the continental shelf to the abyssal plains was the site of major glaciomarine gravity-flow deposits, formed by debris flows, slumps and turbidity currents [THA 05]. These periods of maximum glaciation had profound consequences for Antarctic marine communities and their evolution.

Figure 3.4. *Aerial view of glaciers and ice shelves around the Antarctic Peninsula. During major glaciation phases in the Pliocene, ice shelves could have reached the outer edge of the continental shelf © Thomas Saucède*

4

Southern Ocean Biogeography and Communities

Biogeography is the study of the geographic distribution of species and, by extension, of the ecological (distribution and functioning of communities and ecosystems), taxonomic (richness and composition of faunal collections) and genetic (distribution and relationships between populations) dimensions of biodiversity. A biogeographer's primary objective is to highlight and explain the major distribution patterns of biodiversity. Traditionally, some scientists focus on ecological factors (climate effects and ecological traits specific to species or communities) to explain the distribution of species, while others focus more on the historical aspects of biogeography, that is to say the imprint left by Earth's history and the evolution of species on biodiversity distribution patterns. For example, the presence of marsupial mammals in Australia and South America cannot be explained without taking into account the geographical isolation of these two sub-continents over millions of years, as some part of the evolution of marsupials occurred separately from that of placental mammals. In fact, ecological factors and Earth's history combine to determine the evolutionary history of species. Today, ecological and historical approaches are readily combined, in what is termed a macroecological approach, at least at the scale of the past tens of thousands of years, over the major geographical regions [BRO 95, BRI 07]. Thus, "modern" biogeographers seek to identify, measure and analyze the distribution of biodiversity, as well as the historical and environmental factors that drive it. They also seek to describe and model potential shifts in this distribution to fit expected environmental changes.

Biogeography is of great interest in light of current environmental changes and ever-increasing anthropogenic pressure. It allows us to identify areas of high biodiversity (biodiversity hotspots), to establish ecological monitoring and to propose scenarios about the future of biodiversity. By seeking to identify the environmental parameters responsible for the distribution patterns of biodiversity, biogeography allows us to estimate the potential vulnerability to change of species and communities. This information is vital in the setting up of adequate conservation strategies and programs. For example, the definition of marine protected areas must be based on scientific criteria proven to be ecologically pertinent, and cannot simply be dictated by the economic and political expediency of the moment.

4.1. Inventorying Antarctic marine biodiversity

The inventory of the first Antarctic marine species dates back to James Cook's voyages (1772-1775) and the early whaling and sealing ships of the 18th Century (see Chapter 1). Our knowledge has increased steadily since then through the efforts of the many scientific expeditions that have periodically explored the region since the mid-19th Century (see Figure 1.6). Today, the inventory of Antarctic biodiversity still continues, as the main progress in knowledge so far has been a better understanding of the ecology of species and the functioning of ecosystems [GRI 10].

Over the last 30 years, a revolution in the tools and methods used in molecular analysis has led to significant progress in our knowledge of Antarctic biodiversity, particularly its microbial diversity [KAI 13]. Traditionally, organisms collected would be preserved in formalin, for purposes of histology and systematics. However, this method of preserving a specimen prevented subsequent DNA analyses. From 1990 onwards, organisms have more often been frozen or preserved in ethanol. It is therefore possible to use their DNA and analyze certain genes, such as the mitochondrial gene COI (cytochrome c oxidase subunit I). These molecular genetic techniques have been used to revise the systematics, and to review and reassess the diversity of Antarctic species. They have revealed the existence of previously unknown species and also shown that many marine invertebrate species, previously considered to be separate species, were actually part of a group of cryptic species, indistinguishable morphologically, but genetically different. Conversely, other genetic analyses have demonstrated the existence of polymorphic species that are in fact only one species despite significant morphological differences [DIA 11].

Our knowledge of Antarctic biodiversity has benefited from the development of new tools for mathematical and statistical analysis, as well as digital programs to plot the distribution of species and model the functioning of ecosystems. The establishment of a common database available to all, and the widespread use of geographic information systems that associate each item of information to a precise geographic location (known as georeferenced data) today allow biogeographic information to be summarized and visualized with ease.

4.2. Southern Ocean biogeography

4.2.1. *A rich ocean*

Polar seas have long been considered regions of low biodiversity, no doubt due to the extreme environmental conditions there. However, species richness in the Southern Ocean contrasts strongly with the poverty of the Arctic [CLA 08]. For decades, inventories of species have shown that life in the Antarctic is relatively rich and diverse. These results were confirmed by recent efforts led by the Census of Antarctic Marine Life (CAML) and the International Polar Year (IPY) projects to estimate Antarctic marine biodiversity, in particular that of areas not yet explored, such as the Amundsen Sea, and deep bathyal and abyssal areas. The Southern Ocean as a whole is home to about 5% of the world's marine biodiversity, for a surface area equal to about 8% of ocean surfaces [GRI 10]. Biodiversity on the Antarctic continental shelf greatly exceeds that of the Arctic, but although the Antarctic shelf is vast (11% of the total area of continental shelves), it contains only 8 to 12% of global species richness [CLA 08]. These comparisons could, however, be misleading, since the correlation between species richness and surface area is not linear, meaning that the expected richness for large regions tends to be overestimated [CLA 05]. In addition, this richness must be evaluated in relation to the Southern Ocean's physiographic characteristics. The Antarctic continental shelf is deep and located mainly below the photic zone, and these factors do not favor high biodiversity. The Antarctic coastline represents only 4% of the world's coastal zones, which are the main reservoirs of marine biodiversity. Finally, the diversity of Antarctic coastal habitats is very low, due to the presence of ice and the virtual absence of continental inputs. For a better visual representation of the scale of marine biodiversity in the Antarctic, it should be stated that a comparable level of richness can be found on the Antarctic continental shelf and on the North-West European shelf. And somewhat unexpectedly, this richness is at a similar level to that found in shallow tropical environments, apart from coral reefs.

4.2.2. *Unique biodiversity*

We have seen that the Southern Ocean plays a crucial role in ocean circulation and the planet's climate. It is equally essential for the biosphere, as much for its richness as for the originality of the forms of life it shelters (see Figure 4.1). In the Antarctic, the level of diversity varies greatly from one taxonomic group to another. Antarctic biodiversity can sometimes appear atypical and even unbalanced, in relation to other oceans, as some groups of organisms are under-represented here, such as gastropods, bivalves, decapod crustaceans and fish. In contrast, other organisms proliferate and have diversified to an amazing extent, e.g. pycnogonids, polychaete worms and echinoderms, especially sea urchins, starfish and brittle stars.

The latter two may also be particularly abundant locally, as is the case in the Weddell Sea.

Figure 4.1. *Examples of Southern Ocean marine biodiversity. From left to right and then from top to bottom: Weddell seals, Gentoo penguins, a sea cucumber, a notothenioid icefish, sessile benthic fauna (sponges, sea squirts and cnidarians), a cidaroid sea urchin and a brittle star © Thomas Saucède*

Durophagous predators (i.e. shell-crushers), typically found in the shallow waters of other oceans, are notably absent from the Southern Ocean. True crabs (brachyuran crabs), lobsters, sharks, rays and most teleosts (bony fish) are absent or under-represented. Only a dozen or so species of shrimp and a few anomuran crabs (royal crabs) are to be found in the depths, which is not many considering the size of the ocean. This virtual absence of decapod crustaceans can partly be explained for physiological reasons. At low temperatures, decapod crustaceans are effectively incapable of maintaining ionic balance in their cells. Other more complex reasons could also contribute to the absence of durophagous predators [CLA 08]. There is also a great imbalance in fish faunal diversity, compared to other oceans. A few species of ray are present and there are even fewer sharks, found only in deep water or in sub-Antarctic areas. Only two groups of bony teleost fish are found here: notothenioids (cod and Antarctic toothfish), with the greatest diversity on the Antarctic continental shelf, and liparids, which live in the depths.

The composition of Southern Ocean communities and the functioning of its pelagic and benthic ecosystems are just as unusual. In the pelagic domain, biodiversity is dominated in terms of biomass by a small shrimp, the Antarctic krill, mainly represented by the species *Euphausia superba*. Locally, other organisms can take over: there are salp colonies near the Polar Front, with large copepods in coastal waters.

The most original aspect of the Antarctic pelagic domain is without a doubt its cryopelagic ecosystem, a unique association of living beings, ice and ocean. This ecosystem is also at the center of exchanges between pelagic and benthic domains. Phytoplankton microalgae are the basis of this ecosystem. They live and develop under the ice in association with very diverse microfauna. In winter, when the ice forms, the biomass composed of this microflora and its associated microfauna is considerable. In spring, when the ice melts, this biomass is suddenly released into the water column. This triggers an explosion of life, or a planktonic *bloom*, which ensures the survival and development of a wide variety of living things. A small proportion of this biomass also flows towards the bottom to fuel the benthic communities. Antarctic krill is the main consumer of microalgae. During the austral summer, krill forms huge swarms, which in turn provide a source of food for many other consumers, fish, squid, penguins, and also some seals and species of whales, themselves prey to the ultimate predators: killer whales and leopard seals.

On the seabed, the situation is more contrasted. Poorly diversified benthic communities, mainly composed of deposit-feeding organisms, such as echinoderms (sea cucumbers, brittle stars and sea urchins) and errant polychaete worms, are in marked constrast with communities composed of an abundant, diversified sessile fauna. Dense and vertically well structured, this fixed fauna erects natural, three-dimensional, cathedral-like structures on the seabed, reminiscent of certain tropical reef habitats. The main engineer organisms in these communities are sponges, sea fans, sea anemones, hydrozoans, ascidians and bryozoans, which provide a multitude of micro-habitats for other organisms: amphipod and isopod crustaceans, pycnogonids, errant polychaetes, mollusks, sea cucumbers and other suspension-feeding echinoderms, such as brittle stars and crinoids [CLA 04].

Predators in Antarctic benthic communities are characterized by various feeding behaviors, from carnivorous to scavenging, like ribbon worms or buccinoid gastropods, which thus have a wider choice at their disposal, between live or dead prey. The main benthic predators that feed on fixed or relatively static prey are themselves not very mobile. Amongst them are echinoderms, such as the *Astrotoma agassizii* brittle star species and the *Labidiaster annulatus* starfish, as well as hydrozoan colonies, like *Tubularia ralphii*.

4.2.3. *Richness and latitude*

One of the most striking biogeographic phenomena on our planet is the distribution of its biodiversity along a latitudinal richness gradient, which decreases from the equator to the poles. This gradient is especially relevant to species richness. It is common to all types of organisms, from microbes to plants and animals, and applies to all domains, terrestrial, marine and freshwater [LOM 06]. The latitudinal gradient of richness is ancient, as shown by the distribution of many marine fossils, brachiopods, corals, foraminifera and diatoms. It has existed for at least 250 million years and has even intensified over time, since the beginning of the Eocene Epoch, 50 million years ago (50 Ma). The reasons for this gradient are related to the climatic and oceanographic evolution of the planet (see Chapter 3), and in particular to the global cooling of oceans, the establishment of latitudinal oceanic circulation, the intensification of latitudinal temperature gradients, the development of glacial conditions at the poles, and the shrinking of tropical zones [CRA 04].

The study of the latitudinal richness gradient has generated intense discussions between biogeographers and ecologists over the last century. These studies have led to a better description of the gradient, by demonstrating the existence of a pronounced asymmetry between the marine domains of the Northern and Southern hemispheres. Species richness is in fact stronger in the Southern Hemisphere, especially in the Southern Ocean in comparison to the Arctic Ocean [CRA 04]. To this N–S asymmetry is added an equally marked E–W asymmetry in the Southern Hemisphere. This has been demonstrated for many groups, such as bivalves, gastropods, bryozoans and sea urchins. Richness is stronger in the Indo-West Pacific and south of Tasmania and New Zealand, but weaker in the Magellanic Region. Within the Southern Ocean, south of the Polar Front, however, no such E–W gradient has been observed. This is explained by the homogenizing effect of the ACC, by the uniformity of physical characteristics and water temperature, as well as by the E–W orientation of the continental shelf and the Antarctic coastline [GRI 09]. The absence of a gradient is also explained historically, as the Southern Ocean has long been a distinct biogeographic region, very different from the Magellanic and Indo-West Pacific regions.

One group, however, is an exception, increasing in richness towards the south, with a reverse pattern to other organisms. These pycnogonids are a group of marine arthropods which present extreme diversity in the Southern Ocean compared to other regions. Other cases of reversed latitudinal gradient are only valid for comparisons between the Antarctic and South America. This is the case for certain echinoderms, like sea urchins (see Figure 4.2). Their specific and generic richness is strongest on the Antarctic continental shelf, forming a latitudinal richness gradient that increases between 60 and 70° S.

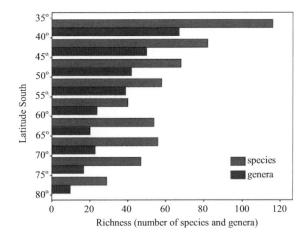

Figure 4.2. *Latitudinal richness gradients of sea urchins in the Southern Ocean, showing an overall decrease between 35° and 80° S, and a slight increase between 60° and 70° S*

4.2.4. Biogeographic regions and provinces

4.2.4.1. Defining some terms

A bioregion, or a biogeographic region, denotes an ecologically homogeneous geographical area – relative to other adjacent geographical areas – from the point of view of its general environmental conditions, and of the organisms living there. Bioregions therefore form geographical units used to describe the distribution and organization of biodiversity on Earth. Each bioregion is characterized by a certain degree of endemism, i.e. with a certain proportion of species present there and nowhere else. The species assemblage unique to a bioregion distinguishes it from other bioregions more or less markedly. Species endemism is both the legacy of the past and evidence of present environmental conditions. The identification of bioregions, or bioregionalization, makes it possible to study the geographical structure of biodiversity in relation to its environment, to characterize relationships between bioregions (known as connectivity), and to estimate the vulnerability of species and ecosystems to environmental change. The identification of bioregions is therefore a necessary prerequisite to establishing marine biodiversity conservation and protection strategies, e.g. the definition of marine protected areas.

Bioregional endemism can be considered at different taxonomic and geographic scales, as families have greater distribution areas than genera, which in turn have greater distribution areas than the species composing them. This double taxonomic and geographic hierarchy has led to the identification of several biogeographic regions on Earth. These regions have next been subdivided into sub-regions, which are in turn composed of provinces, and then of districts, along a spatial scale decreasing in size [LOM 06]. With regard to the Southern Ocean, various approaches have been used for regionalization. They differ mainly by the way in which faunal similarity and environmental parameters are taken into account. Some biogeographers include only geographic and physical criteria, which they consider to be sufficiently structuring and representative, but exclude biodiversity, which is seen as too complex and poorly known to be directly included in analyses [DOU 14]. Conversely, others use a more or less complete and detailed composition of the fauna and flora present in different sectors, without directly taking into account the environment [GRI 09]. One final, still emerging trend attempts to analyze the distribution of species and the biological and physical factors in their environment with an integrated approach, often restricted to specific sectors of the ocean [KOU 10]. This approach is properly termed ecoregionalization, as it seeks to identify ecoregions in the ecological sense of the term, i.e. regions that are homogeneous in terms of environment and biodiversity.

4.2.4.2. Regions of the Southern Ocean

While studying the distribution of mammalian species on the surface of the globe, Allen [ALL 78] was the first to define an Antarctic "zoogeographic" region. Following him, other scientists continued to refine the biogeographic division of the Southern Ocean, based on the inventory and distribution of a greater or smaller number of different organisms: species of fish alone [GIL 84], benthic and pelagic species [ORT 96], echinoderms and fish [EKM 53] or organisms from the photic zone [KNO 60]. They integrated into their work, either directly or indirectly, various physical environmental parameters, such as temperature or other more complete hydrographic and oceanographic characteristics. In terms of the quantity and diversity of organisms analyzed, the first true biogeographic synthesis of the Southern Ocean was produced by Hedgpeth [HED 69]. This study, until recently the benchmark reference, took into account the distribution of species from 21 groups of marine benthic and pelagic invertebrates, but without including any information about their environment. This bioregionalization of the Southern Ocean recognized two large regions: sub-Antarctic and Antarctic, which were subdivided into several sub-regions and districts: the Magellanic Sub-region, Tristan da Cunha District, and Kerguelen Sub-region, and also the High Antarctic Region, Scotia Sub-region and South Georgia District (see Figure 4.3).

In recent decades, growing concern for the protection and conservation of biodiversity, confronted with the dangers of environmental change, has generated among biologists a renewed interest in biogeographic studies. Many works have since been published on marine invertebrates, plankton, fish and Antarctic predators.

They rely on the development of new calculations and tools for analysis. While following the broad outline of the framework defined by Hedgpeth, biogeographers today highlight the difficulty of defining bioregions valid for all types of organisms, whatever their depth range, including isolated oceanic islands inhabited by endemic fauna. As an example, the differentiation between East and West Antarctic may be relevant for certain organisms, like gastropods, but not for others, like sea urchins, bivalves, pycnogonids, bryozoans and sponges. The same applies to the faunal relationships between South America, the sub-Antarctic islands (South Georgia, Bouvet, Marion, Prince Edward, Crozet, Kerguelen and Macquarie) and New Zealand. Some biogeographic studies have therefore targeted specific ecosystems, at specific depth ranges [BRA 07], for certain regions and islands [BAR 08]. To give an example, several categories of benthic provinces have been recognized by UNESCO [UNE 09], according to depth range. It has thus been proposed that the bathyal zone of the Southern Ocean (200 to 2,000 m) could be subdivided into two latitudinal provinces, sub-Antarctic and Antarctic, while the abyssal plains (2,000 to 4,000 m) could be structured as two longitudinal provinces: East and West Antarctic [PIE 13].

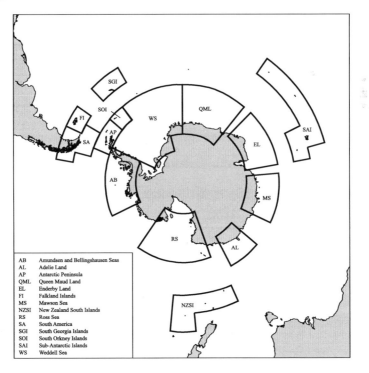

Figure 4.3. *Biogeographic sub-regions and districts of the Southern Ocean, established from [HED 69], [LIN 06] and [GRI 09]. Redrawn from [PIE 13]*

Recent studies by biogeographers have agreed on the definition of two major regions in the Southern Ocean: the Antarctic Region south of the Polar Front, and the sub-Antarctic Region to the north [DEB 14]. In the Antarctic Region, most benthic species are distributed more or less evenly over the entire Antarctic continental shelf. The difference between east and west is more or less marked depending on the organism studied. Within the Antarctic Region, South Georgia Island is often considered as a separate district. In the sub-Antarctic Region, the southern tip of South America (Magellanic Sub-region) is often singled out by biogeographers. Finally, as the sub-Antarctic islands are under the dual influence of South America to the west and New Zealand to the east, they are grouped differently, depending on the organisms studied. Despite these subtle differences, all biogeographers agree on the existence of general faunal continuity between the regions and provinces of the Southern Ocean. This network of relationships is attributed to the structuring effects of the ACC and the Antarctic Coastal Current [PIE 13] (see Figure 4.4). It is also explained by the legacy of climatic and oceanographic history, as exemplified by marine trans-Antarctic connections that no longer exist, but which can explain the similarities between the Atlantic and Pacific zones of Antarctica, separated today by the continent (see Figure 4.5).

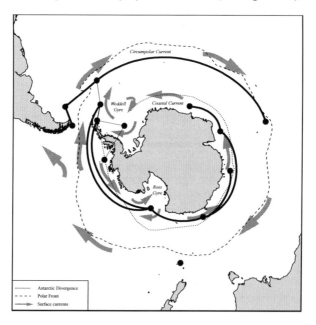

Figure 4.4. *Continuity of sea urchin fauna between regions of the Southern Ocean. Strong fauna similarities (thin black lines) between regions (black dots) were calculated from species distribution data. Bold black lines indicate the concordance between faunal similarity and surface marine currents (ACC and Antarctic Coastal Current), favoring species dispersal around the Antarctic. Redrawn from [PIE 13]*

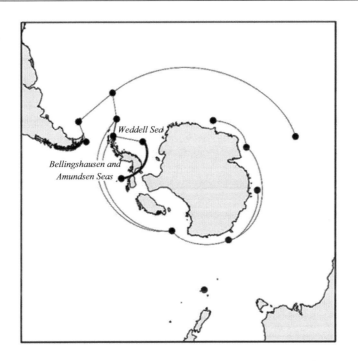

Figure 4.5. *Sea urchin faunal similarity between regions of the Southern Ocean, positioned on a paleogeographic map of the Pleistocene. Strong faunal similarities (thin black lines) between regions (black dots) were calculated from species distribution data. The bold black line indicates sea urchin faunal similarity between the Weddell Sea and the Bellingshausen and Amundsen Seas. This present-day similarity is the legacy of an ancient seaway that existed in the Pleistocene (around one million years ago), between regions separated today by the Antarctic Peninsula (see modern map, Figure A.1 in Appendix). Redrawn from [PIE 13]*

The overwhelming majority of biogeographic work so far has focused on macrofauna and megafauna, i.e. organisms visible to the naked eye. The study of meiofauna, i.e. organisms of millimeter to infra-millimeter size (nematodes, copepods, etc.), has only recently begun in biogeography. Scientific results are nevertheless contrasted, depending on whether the data studied are morphological or molecular. A study based on the morphological identification of ostracod species concluded that strong Antarctic endemism existed, in association with circum-Antarctic homogeneity, while another study, based on genetic analyses, concluded that there was only weak endemism [KAI 13]. No final agreement has yet been reached.

For classes of even smaller size (microfauna), diversity is difficult to measure in terms of abundance or richness. Nevertheless, recent progress has allowed recurrent patterns to be identified. The composition of bacterial colonies in Antarctic marine ecosystems thus appears to result mainly from primary productivity and substrate availability. In contrast, the diversity of these communities seems to be determined by particulate organic matter flux.

Regionalization of the Southern Ocean today is thus largely based on the analysis of similarities between marine fauna from different regions, provinces and districts. Some recent attempts have, however, been made to establish a biogeographic classification of Antarctic ecosystems, i.e. regionalization of the Southern Ocean based on criteria related to both oceanography and ecology, an aspect which is often overlooked in biogeography. Thus, both biogeochemical and oceanographic data have been used to subdivide the Southern Ocean into six provinces: Austral Polar, Antarctic, Subantarctic, Humboldt Current Coastal, SW Atlantic Shelves and New Zealand Coastal [LON 07]. Other works have sought to integrate notions related to ecosystemic services, or have even included faunal ecological and evolutionary traits for conservation purposes. The system used in *Marine Ecoregions of the World* [SPA 07] is based on faunal similarity analysis that integrates evolutionary, dispersal and isolation criteria. This system has identified five provinces in the Southern Ocean: Temperate South America (Magellanic Province), Sub-Antarctic Islands, Scotia Sea, Continental High Antarctic and Sub-Antarctic New Zealand.

To conclude, whatever approach was used, these different proposals for Southern Ocean regionalization all agree on a basic pattern, the spatial distribution of austral biodiversity in four regions or provinces: Antarctic, Sub-Antarctic, South American and New Zealand.

4.2.4.3. *Endemism and connectivity*

The identification of distinct biogeographic regions implies the existence of a certain degree of species endemism within these regions. In other words, each region is characterized by a species assemblage that distinguishes it from neighboring regions. These peculiarities stem from the prevalence of environmental conditions characteristic to each region. They also stem from the existence of more or less ancient and permeable geographic and oceanographic barriers that isolate the species of different regions, with endemism tending to increase with age and isolation. This is the case for the Southern Ocean as a whole, as it has been isolated for millions of years (see Chapter 3, section 3.1).

Although endemism was thought to be very high in the Antarctic, from 75% to 90% of marine invertebrate species, this figure has recently been revised downward. It has also been shown that endemism is very variable, depending on the taxonomic groups considered. The highest levels of endemism are, without contest, observed in benthic marine invertebrates. Endemism can exceed 70% for gastropods, pycnogonids and some crustaceans [ARN 97, DEB 96]. It is less marked in cephalopods, bivalves, sea urchins and bryozoans. Conversely, the lowest levels of endemism are found in pelagic and planktonic species. Pelagic species, which are active swimmers, can of course disperse latitudinally, and thus overcome obstacles, such as the ACC and the Polar Front. The high level of endemism observed in benthic species is explained by the ancient isolation of the Antarctic continental shelf, separated not only oceanographically but also geographically from other continents by deep ocean basins extending over thousands of kilometers. This ancient isolation was accompanied by strong endemism in groups that include many species with direct development, without dispersive larval stages. Most of them brood their young and have limited dispersal ability (see Chapter 6). Such differences in endemism, and therefore in connectivity, have very real consequences on the vulnerability of a species, and the most isolated species, with the most limited distribution area, are the most fragile when confronted with environmental change [GRI 09]. Spatially, the Weddell Sea is the region with the highest level of endemism. This is explained by the local absence of the Antarctic Coastal Current and by the presence of a strong gyre that contributes to the isolation of the sector. The sub-Antarctic islands in the South Atlantic, Prince Edward and Bouvet, also show strong endemism.

Endemism is generally found where there is little or no connectivity between regions. Thus, faunal similarity between New Zealand and Antarctica is very low. In contrast, faunal similarity is high between the Magellanic and Peninsular regions, despite the presence of the Polar Front and the ACC between the two continents [DEL 72]. This singular faunal link is explained by transitory exchanges resulting from fluctuations in the position and intensity of the ACC in the distant past. Biogeographers think that South American fauna may have entered the Southern Ocean on several occasions over the past few million years via the islands of the Scotia Arc and the sub-Antarctic islands. Species then dispersed to the east via the ACC. It is thought that some species are still in the process of dispersal over the Antarctic continental shelf. Indeed, many species forced from the Antarctic shelf by advancing ice sheets during the Last Glacial Maximum could have found refuge further north in sub-Antarctic regions and the Scotia Arc. With the retreat of the ice, these species gradually recolonized the Antarctic continental shelf, a process that could take several million years for species with limited dispersal ability. Connectivity between different sectors of the continental shelf is favored by its continuity and the presence of dispersal vectors such as marine currents.

Some connectivity also exists within the sub-Antarctic region. In addition, the sub-Antarctic islands of Crozet, Kerguelen and Prince Edward share many species with the Magellanic Region or, at least, closely related species can be found there. Once more, this connectivity is provided by the dispersive role of the ACC. Indeed,

South American influence on the sub-Antarctic islands gradually diminishes toward the east, in the same direction as the flow of the ACC.

4.2.5. *The paradox of bipolar distribution patterns*

Several closely related (genus and family) species, or groups of species, have extensive distribution areas and are present both in Arctic and Antarctic polar waters [STE 06]. This is termed bipolar distribution and concerns almost 200 species, and more than 800 genera and 600 families, composed respectively of sister species and closely related genera [DEB 14]. Amongst species with bipolar distribution patterns are single-celled organisms, dinoflagellates and radiolarians, but also invertebrates, such as hydrozoans, jellyfish, sea anemones, polychaete worms, bryozoans, copepods, pteropod gastropods and some echinoderms.

Some species with bipolar distribution patterns are absent from lower latitudes (temperate and tropical waters) and therefore show a disjunct distribution pattern. Others are clearly cosmopolitan, i.e. they have been found at all latitudes. Whether their distribution is disjunct or not, bipolar species have long aroused the curiosity of marine biologists. How can the same species maintain connectivity and genetic exchanges over such vast areas, especially when they are disjunct? And, in the case of cosmopolitan species, how can they survive in such contrasting latitudes and environmental conditions? Three explanations pertaining to different taxonomic levels, at different time scales, have been put forward:

1) The first involves the establishment of a biogeographic barrier that divided an initial single distribution area into two disjunct parts. Thus the break-up of the supercontinent Pangaea, that united all landmasses until the Jurassic, could explain the separation of vast distribution areas of ancient marine coastal species, and their slow drift towards the poles during the Mesozoic. This scenario could explain the bipolarity of some families of marine invertebrates [CRA 93]. It is a plausible explanation for divergence between ancient taxa.

2) Another scenario suggests that the warming of tropical waters in the Miocene could have progressively affected the dispersal of temperate and cold-water species between the two hemispheres by pushing them towards cooler waters in high latitudes.

3) Finally, recent glaciation cycles explain the bipolar distribution of many extant marine species of mollusks, fish and crabs. As tropical waters cooled during glacial maxima, the dispersal of populations between the two hemispheres was facilitated. In contrast, the warming of tropical waters during interglacials limited the dispersal of populations, leading to disjunct distribution areas. According to this scenario, most disjunct distribution must therefore be fairly recent, as there has not been enough time since the Last Glacial Maximum, around 20,000 years ago, for the

appearance of new species through isolation (the phenomenon of vicariance). Bipolarity would therefore be merely a transitory biogeographic phenomenon [PAR 09]. Scientists prefer this scenario as an explanation of bipolar distribution in extant species.

In the case of cosmopolitan species, another hypothesis has been put forward. The maintenance of gene flow and cohesion between northern and southern populations could be possible because of the equatorial submergence phenomenon, facilitated by downwelling, as cold waters plunge deeper in low latitudes. Bipolar and cosmopolitan species could thus survive in all latitudes, without suffering from too great a contrast in environmental conditions. They would simply follow the circulation of deep cold waters upwelling in Arctic and Antarctic polar regions [STE 06]. The cold waters of the Pacific Coastal Current that flow along the western coastlines of America could also ensure environmental continuity between polar regions [PEA 04].

5

History of Biodiversity in the Southern Ocean

The Southern Ocean is home to biodiversity that is, in many ways, unique. Some groups of organisms, such as decapod crustaceans, or cartilaginous and teleost fish, are vastly underrepresented there. In contrast, other groups, such as isopod and amphipod crustaceans, echinoderms, ascidians, pycnogonids and polychaetes, proliferate there, with unusual forms of adaptation. The composition of marine communities is so atypical that some scientists have even suggested that they could be relict communities from the Paleozoic Era [ARO 07]. What are the origins of marine biodiversity in the Southern Ocean and when did it appear? Where does its uniqueness come from? Could very ancient biodiversity have been preserved, to evolve separately from other oceans around the world? What influence did the many changes in climate have on its evolution and, in particular, what were the consequences of the many successive phases of intense glaciation, over millions of years?

The evolutionary history of Antarctic marine fauna and flora is closely linked to the tectonic, oceanographic and climatic events that have punctuated the history of the Southern Ocean. Few outcrops are accessible on the margins of this ice-covered continent. Fossils are rare and mainly found in deposits more recent than the Mesozoic Era. Valuable additional information is obtained from modern genetic analysis. This historical evidence shows that the major geodynamic events that have occurred on the planet from the Cretaceous Period onward have profoundly marked the evolution of biodiversity in the Southern Ocean, leading either to the extinction or the diversification of many species.

5.1. So much ice yet so few fossils

Fossils provide evidence of ancient biodiversity and its evolution. Unfortunately, such evidence is rarely accessible in the Antarctic. The thick ice sheets that almost entirely cover the Antarctic continent have sealed off fossiliferous outcrops, thus closing a virtual window into ancient worlds, at least on the timescale of the human lifespan. Fortunately, the absence of ice in some areas has uncovered some exceptional fossil-bearing beds, discovered and studied by paleontologists, in some instances almost a century ago. The invaluable data gained from these rare sites provide descriptive evidence of ancient biodiversity in one area of the Southern Ocean, allowing us to retrace some part of its history.

Our knowledge of ancient Antarctic biodiversity is therefore incomplete. It is also very heterogeneous. Antarctic paleontology and paleobiogeography is mainly based on the study of some groups of marine vertebrates and invertebrates, such as bivalves, gastropods, echinoderms and brachiopods [SAU 13, REG 14]. These fossils are from certain geological periods only, mainly towards the end of the Cretaceous Period (end of the Mesozoic Era), the beginning of the Paleocene Epoch (beginning of the Cenozoic Era), and in the interval between the end of the Eocene and the early Miocene. Nevertheless, these few markers suffice to give a relatively good idea of the history of Antarctic biodiversity, allowing scientists to study the consequences of major geodynamic, climatic and oceanographic phenomena on the evolution of Antarctic fauna and flora. Fossil evidence provides information about the Cretaceous/Tertiary (K-T boundary) mass extinction event, at the end of the Mesozoic Era, the break-up of the supercontinent, Gondwana, and the establishment of the Antarctic Circumpolar Current (ACC), with its consequences on the evolution of fauna and flora on the Antarctic continental shelf.

The rare fossil remains found came mainly from the west of the continent and along the Antarctic Peninsula, from three islands, Alexander, James Ross and Snow Hill, for the Cretaceous, from King George Island for the Oligocene and the early Miocene, and from Seymour Island for the end of the Cretaceous and the Eocene. Fossil beds in East Antarctica also yielded a small quantity of fossils, for example from Black Island and the McMurdo Sound, for the end of the Eocene. The geological La Meseta Formation, on Seymour Island, is definitely the most fossiliferous in the Antarctic. Dating from the Eocene, many fossils found there suggest the existence of an ancient biodiversity rich in marine vertebrates and invertebrates. This formation has yielded cartilaginous (sharks, sawfish and rays) and bony (teleost) fish fossils, decapod crustacean fossils (crabs and lobsters) and echinoderm fossils. The composition of the fossil fauna found there is very different from extant fauna. In particular, some key groups in coastal ecosystems, such as large carnivorous predators, were well represented then, although they are now extinct in the Antarctic (see Chapter 4 section 4.2.2). Some rare fossils show the partial persistence of such fauna after the Eocene Epoch. Examples are a species of crab discovered in marine sediments from the Miocene, and a species of lobster from the Pliocene. These indicators show that the current absence of large decapod crustaceans in the Antarctic does not correspond to complete extinction at the end of the Eocene, but is rather the result of a lengthy process, spanning tens of millions of years [CLA 04].

5.2. Origins and age of Antarctic marine biodiversity

Antarctic marine fauna greatly astonished the first explorers, and its originality continues to fascinate, with the question of its origin still causing much debate amongst biologists and paleontologists today [KNO 77]. Over the past five million years, sea-level variation cycles and the succession of ice-sheet advances and retreats over the continental shelf have greatly disrupted marine habitats. These phenomena were not, however, sufficient to eradicate the marine communities living there, even during glacial maxima. The Antarctic continental shelf has therefore been inhabited since ancient times by most of the groups of organisms still living there today [CLA 89]. The absence of large predators usually present in marine ecosystems and the predominance of filter feeders (sponges and sea squirts) on the seabed confer on the marine community an "archaic" character, which for some scientists recalls the Paleozoic communities that were once widespread throughout all oceans. It is nevertheless recognized today that Antarctic marine communities are more recent in origin, and that the absence of large predators and the predominance of sessile fauna is an evolutionary answer to the glaciomarine conditions progressively established during the Cenozoic Era, rather than a legacy of the distant past.

There are two main scenarios to explain the origins of Antarctic marine biodiversity. One applies to shallow marine fauna, the other to deep fauna:

(1) According to the first scenario, known as *Out of the Tropics*, many species present today in high latitudes originated from tropical lineages. In this hypothesis, the tropics represent "source" regions for marine biodiversity that regularly supply higher latitudes with new taxa. The Indo-West Pacific and Caribbean zones would thus be sources of biodiversity for the Southern Ocean, which would become a repository for new lineages (see Figure 5.1). This scenario is supported by paleontological data confirming the existence in the tropics of a greater proportion of taxa that are young in evolutionary terms, and therefore greater diversification (evolution of new species) in tropical regions than elsewhere [FLE 96]. The study of bivalve fauna has also shown that the first known representatives of many groups had mainly a tropical distribution, and that the average age of lineages increases toward the higher latitudes [JAB 06].

(2) A different scenario has been proposed to explain the origin of deep marine species. According to this second scenario, the *Thermohaline Expressway*, Antarctic fauna supplied the deep ocean basins of lower latitudes through the phenomenon of polar submergence [STR 08]. In Chapter 2, we saw that Antarctic bottom waters flowed north, feeding into global thermohaline circulation. These water mass movements encourage the dispersal of Antarctic species toward deep ocean basins further north. This scenario is supported by studies based on a variety of marine invertebrate groups, including isopods, octopuses, foraminifera, sea cucumbers, polychaetes and sponges.

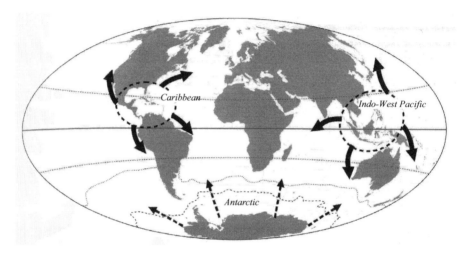

Figure 5.1. *The three source regions for marine biodiversity. According to the Out of the Tropics scenario, Caribbean and Indo-West Pacific are the source regions for high-latitude marine biodiversity, in particular for the Southern Ocean. According to the Thermohaline Expressway scenario, the Antarctic continental shelf is the source region for low-latitude deep marine biodiversity. From [DEB 14]*

Other hypotheses specific to the Southern Ocean have also been proposed to explain the origin of Antarctic marine biodiversity. None is exclusive, as Antarctic marine biodiversity has at least three origins [KNO 77]:

(1) Part of the Antarctic fauna could in fact be indigenous, i.e. it has been present in the Antarctic since the end of the Mesozoic Era, evolving there since then. Some families of marine invertebrates have a long evolutionary history in the Antarctic that can be traced back to the Cretaceous Period, as has been confirmed for two families of gastropods: Struthiolariidae and Buccinidae [CLA 10].

(2) According to another hypothesis, part of the fauna originated from the deep ocean basins surrounding the Southern Ocean in a process known as polar

emergence. This situation, seen in some isopod crustaceans, is the exact opposite of the *Thermohaline Expressway* described earlier [DEB 14]. Yet these two propositions are not incompatible. In fact, certain species may have colonized the Antarctic continental shelf from adjacent deep ocean basins, while some species of Antarctic fauna migrated in the opposite direction toward the deep basins. These dual faunal exchanges would have been facilitated because the temperature difference between the cold Antarctic waters and the deep ocean was less than 4°C, even if the difference may have been a little higher in the past [ZAC 01].

(3) Finally, according to a third hypothesis, part of Antarctic fauna could be South American in origin, and may have spread south via the islands of the Scotia Arc, while some Antarctic species may have migrated north along the same path.

In conclusion, many scenarios have been proposed to explain the origins and antiquity of marine biodiversity in the Southern Ocean. What must be remembered is that the uniqueness of current Antarctic biodiversity is the result of lengthy evolution *in situ*, but also the result of many faunal exchanges with lower latitudes, with different itineraries for shallow and for deep fauna.

5.3. Break-up of Gondwana and isolation of Antarctic fauna

At the end of the Mesozoic and the beginning of the Cenozoic Era, the break-up of the supercontinent Gondwana had far-reaching consequences on the evolution of marine biodiversity in the Southern Hemisphere. Throughout the Mesozoic Era, marine fauna in high southern latitudes had adapted to life in warm, shallow seas and shared many affinities with fauna of lower latitudes [CRA 04]. In the Cretaceous, these southern fauna, cosmopolitan until then, began to differentiate from fauna in other oceans [CLA 89]. The respective movements of Africa, South America and Antarctica led to the gradual isolation of Antarctic fauna, but also to convergence between the Antarctic Peninsula and Patagonia, induced by the formation of the Weddellian Isthmus, which promoted exchanges between the two continents (see Chapter 3). Endemism in southern fauna then became more marked, and a large biogeographic province came into existence, the Weddellian Province, stretching along the southern margins of Gondwana, from Patagonia to New Zealand, and to the southeast of modern-day Australia via West Antarctica [ZIN 79].

At the beginning of the Cenozoic Era, the remaining parts of Gondwana continued to separate and the fauna of the Weddellian Province began to evolve separately. New Zealand and Australia began their northward drift to warmer climates. Their fauna therefore evolved, differentiating from Antarctic fauna. After the flooding of the Weddellian Isthmus and the formation of the Drake Passage, Patagonian and Antarctic species began, in their turn, to differentiate [DEB 14]. In contrast, the establishment of South Pacific surface currents, and then of the ACC, facilitated the dispersal of species between South America and New Zealand, from the Oligocene to the Pliocene [SAU 13].

The consequences of the ancient break-up of Gondwana on southern fauna can still be observed today in the composition of many groups of marine organisms. For example, notothenioid fish, a very diversified group on the Antarctic continental shelf today, have a Weddellian origin dating back to the Late Cretaceous. The ancient origins of this group explain the presence of the same fish today in South America, Tasmania and Southern Australia, regions formerly in the Weddellian Province. The species in these regions, however, are different from Antarctic species as they have evolved independently of each other for approximately 40 million years.

5.4. Mass extinction event at the end of the Mesozoic Era

The mass extinction that marked the transition from the Mesozoic to the Cenozoic Era severely affected marine biodiversity. While 70% of species disappeared from the surface of the Earth and its oceans, 50% of genera and 70% of marine species also disappeared from the Antarctic continental shelf [JAB 08]. This extinction is well documented by the many fossil remains of shelled mollusks that have been found. Of the 26 taxa of Antarctic mollusks identified from the Late Cretaceous, only nine still survived at the beginning of the Cenozoic Era [STI 03]. The faunal assemblages also changed. Mollusk fauna from the Cretaceous were dominated by ammonites and suspension-feeding bivalves whereas carnivorous gastropods predominated in the Cenozoic Era. Although the mass extinction in the Late Cretaceous caused a drop in global biodiversity, this decrease was then followed by significant diversification, more marked in the tropics than at the poles, which began in the Eocene and continued throughout the Cenozoic Era [KRU 09]. The wealth of Eocene fossil remains on Seymour Island provides evidence of this great diversification of Antarctic marine fauna at the beginning of the Cenozoic. Many marine groups, today endemic to the Southern Ocean, first appeared at that time: notothenioid fish, penguins, gastropod mollusks, bivalves and corals [BEU 09]. Only one bivalve genus from the Cretaceous, *Malleti*, is still present today, whereas 6 genera and 13 families of bivalves from the Early Cenozoic still exist. Among whelks (neogastropoda), certain genera that first appeared at that time are still present in the Antarctic today, e.g. the genus *Probuccinum* [STI 04].

5.5. Evolution of biodiversity and ancient climatic changes

Sudden events, long-term trends, climatic rhythms and oscillations have strongly affected the evolution of southern marine biodiversity. Despite several global warming events and periods, the evolution of Earth's climate during the Cenozoic Era was mainly marked by an overall cooling trend (see Chapter 3, section 3.1). This long evolution was marked by several upheavals that were not without consequences for Antarctic marine biodiversity. After a temperature peak during the Paleocene-Eocene transition (Paleocene-Eocene Thermal Maximum), three episodes in particular marked this lengthy cooling trend: the first glaciations and the cooling of the Southern Ocean during the Eocene-Oligocene transition, 34 million years ago, the mid-Miocene cooling step, around 15 million years ago, and then successive glacial-interglacial cycles, from around 5 million years ago.

5.5.1. *The Paleocene-Eocene Thermal Maximum*

In the first part of the Cenozoic Era, the Paleocene-Eocene transition, dated 55 million years ago, was marked by an abrupt increase in temperature. During this event, subtropical waters probably reached the coasts of the Antarctic continent. This thermal event, devastating for terrestrial fauna and flora, also affected marine biodiversity, even in deep ocean zones [ZAC 08]. Although up to 50% of benthic foraminiferal species disappeared, there was also significant renewal of planktonic organisms.

5.5.2. *Consequences of the late Eocene biological crisis*

At the Eocene-Oligocene boundary, around 34 million years ago, a major cooling event had profound repercussions on the composition of Antarctic marine fauna, causing partial and even total regional extinctions in many groups: giant penguins, decapod crustaceans, stomatopods, barnacles, teleost fish, cartilaginous fish, bivalves and echinoderms [DEB 14]. These extinctions were very selective, primarily affecting groups unable to adapt to cold water for various reasons, whether physiological (thermoregulatory ability) or ecological (reproductive strategies). Such was the case for durophagous predators, such as teleost and cartilaginous fish, and decapod crustaceans. Other organisms, such as bivalve clams that live buried in sediment, which were still very much present in the Antarctic during the Eocene, have been completely absent since that time, whereas many clam species thrive in South America today. There is not necessarily a direct and simple causal relationship between cooling water masses and extinction of species. However, although most decapod crustaceans were rare in the Antarctic after the Eocene, some species, like the homolodromiid crabs, were still common in the Miocene. Complex ecological factors were also decisive, including food availability and quality, the structure of the food network, and available habitats.

These extinctions were associated with a significant change in the composition of marine communities, well documented through the numerous marine vertebrate and invertebrate fossils found on Seymour Island. Dating from around 40 million years ago, these fossils provide evidence of the onset of water cooling, with the transition from cool to cold conditions [ARO 09]. New communities started to appear, dominated by dense, sessile or relatively immobile fauna composed of echinoderms, brachiopods and bryozoans. The structure of these communities is reminiscent of Palezoic marine communities but they only became established in the late Eocene. The development of these new marine communities can be explained by the establishment of glaciomarine conditions in coastal environments, while the rich coastal biodiversity, inherited from the warm seas of the early Cenozoic Era, disappeared. The great reduction in sedimentation rate, and thus in turbidity, is one of the principal factors leading to the development of communities rich in sessile suspension-feeding organisms. These organisms, often relatively immobile, were also able to prosper following the disappearance of large durophagous predators.

The extinctions that occurred in some groups thus created conditions favorable to the diversification and development of other organisms. In fact, after major extinctions, factors such as lower interspecies competitive pressure, less dense occupation of habitats, together with reduced competition for energy resources, all favored the diversification of organisms that had developed physiological (e.g. antifreeze proteins) or ecological innovations enabling them to prosper in new environments. Periods of extinction and speciation are therefore related. The extinction of decapod crustaceans thus favored the emergence and diversification of peracarid crustaceans (isopods and amphipods). The numerous extinctions that occurred among teleost fish favored the diversification of notothenioid fish. Among gastropods, a similar phenomenon favored the development of species of Buccinidae, Naticidae and Turridae. The end of the Eocene is therefore characterized by diversification in fish, crustaceans, gastropods, bivalves, scaphopods, barnacles, penguins, and many other groups of marine organisms (see Figure 5.2).

The cooling of the Southern Ocean and the appearance of the first ice on the Antarctic continent were conditioned by the thermal isolation of the ocean, linked to the strengthening of the ACC and the opening of the Drake Passage. The dates of divergence between Antarctic and sub-Antarctic species in many groups of marine organisms, such as notothenioid fish, echinoderms, krill, algae, etc., coincide with the age of this isolation, which was both geographic and oceanographic [LEE 04].

In summary, at the end of the Eocene, extinction, diversification and isolation were the three factors that together created the necessary conditions for the evolution of unique marine fauna and flora in the Southern Ocean.

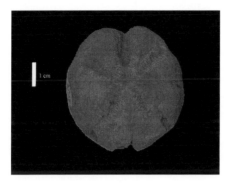

Figure 5.2. Abatus kieri *(Schizasteridae)* an Eocene sea urchin fossil from Seymour Island (collection J. Stilwell, Monash University). *This sea urchin family experienced great diversification after the Eocene and today forms one of the richest groups of Antarctic sea urchins, in terms of species © Thomas Saucède*

5.5.3. *Glaciation and species adaptation in the Miocene Epoch*

A new period of global warming marked the end of the Oligocene, intensifying in the early Miocene, between 23 and 17 million years ago. Unsurprisingly, the early Miocene warm period coincided with a renewal of marine fauna in the Antarctic, as evidenced by the fossils found in the South Shetlands Archipelago. Unusual fauna have been found, i.e. decapod crustaceans and burrowing bivalves, including clams and razor shells, all of which are absent today from the Southern Ocean. These fauna were not present for very long, because another strong cooling event began in the mid-Miocene, around 15 million years ago.

This new cooling event was the beginning of a true glacial phase that greatly modified marine environments. Ice sheets formed first over the East Antarctic, between 14 and 12 million years ago, but only later in the West Antarctic, between 8 and 5 million years ago. This glaciation was accompanied by a sudden drop in primary productivity in the ocean, and therefore a reduction in trophic resources. It also caused the fragmentation and destruction of coastal habitats after the advance of ice shelves over the sea. All these disruptions led to the extinction of species inhabiting shallow environments, composed of small localized populations. This happened to some bivalves now adapted to temperate waters, such as mussels and oysters, to large algae, and to barnacles and limpets that live attached to the rocky seabed. Also affected were species with a planktotrophic larval stage during their growth. As their larvae fed on plankton, the decrease in primary production strongly interfered with their development and survival. Among fish species, fast predators, with a high metabolic rate only possible in warm waters, were unable to survive in the cold Antarctic waters of the Miocene.

However, the decrease in temperature during the mid-Miocene was gradual and did not cause any extinctions. It also left enough time for many species to evolve and adapt to their new environmental conditions, in particular to the cold. This was the case of the notothenioid fish, which developed a mechanism for the production of antifreeze proteins to withstand the cold. This physiological innovation occurred at the same time as the first glacial phases in the Southern Ocean, thus ensuring the evolutionary success of the group.

In the absence of fossils, genetic analysis of extant species enables their evolutionary history to be retraced and is particularly useful in attempts to date the main events in species divergence, and the emergence of new species. Genetic data show that many species probably diverged in the mid-Miocene [GON 10]. The diversification of new species adapted to such cold waters was also promoted by the strengthening of the ACC, which favored the dispersal of populations over large distances around the Antarctic and led to the differentiation of new local species [PEA 09].

5.5.4. *Are glacial-interglacial cycles good for biodiversity?*

After the Miocene glaciation, there was a new period of global warming during the Pliocene, between 5 and 2.5 million years ago. Yet again, this warming event was marked by the return to the Southern Ocean of species from cold temperate waters, such as scallops, decapod crustaceans and dolphins. Fossil remains of these organisms have been found in Pliocene deposits. Such species were able to migrate from neighboring South America via the islands of the Scotia Arc, when climate conditions were suitable. These were the last-known cold temperate water fauna to have lived in the Southern Ocean; they finally disappeared from Antarctic waters around 2 million years ago, when glaciation events intensified [BER 96].

From the Pliocene to the present day, the Southern Ocean's climatic history has been marked by successive glacial-interglacial cycles that have greatly affected and even challenged the survival of marine communities, especially on the continental shelf. Thirty-eight glaciomarine cycles have occurred over the past five million years. Geological data suggest that ice sheets may have occasionally covered the whole of the Antarctic continental shelf, and that slumps and debris flows probably occurred along the continental slope. During glacial maxima, sea ice remained present along the coastline throughout the year, and multi-year sea ice extending far northward limited marine primary production. These glacial conditions were recurrent, reoccurring relatively rapidly, at the scale of species evolution, and most certainly had a catastrophic effect on marine habitats and communities. To what extent did glacial maxima affect marine biodiversity in the Southern Ocean and how have extant Antarctic fauna survived?

Some extant Antarctic species have ancestors identified from fossils dating (very rarely) from the end of the Mesozoic Era, or (more commonly) from the beginning of the Cenozoic Era. Some groups have therefore managed to survive for a considerable length of time in the Antarctic, even through the highest glacial maxima [CLA 10]. This is even the case for shallow-living species, such as the limpet, *Nacella concinna*, the scallop, *Adamussium colbecki*, and the clam, *Laternula elliptica* (see Figure 5.3). These mollusks have existed in the Antarctic since the Pliocene, as evidenced by many fossils. They have therefore survived glacial maxima, perhaps by taking refuge at the top of the continental slope, or by sheltering behind certain large islands (Alexander and Thurston), which may have formed natural barriers, limiting the expansion of ice sheets along ocean-facing coastlines. Geologists point out that certain distal areas of the continental shelf have never been reached by ice (Prydz Bay, Ross Sea, and George V Land). Moreover, even during the strongest glaciation, a significant proportion of glacial flow was channeled by submarine canyons. Finally, the maximum extension of the ice was not synchronous across the whole of Antarctica. Many transitional refuges would have sheltered marine fauna locally along the Antarctic continental shelf, with species recolonizing the seabeds during the interglacial phases. This scenario is corroborated by the broad range of vertical distribution of many Antarctic species, which were able to take refuge at greater depths during ice advance.

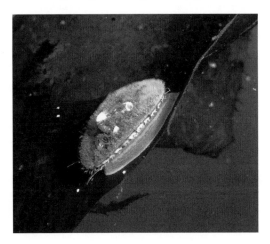

Figure 5.3. *Limpet species of the Nacella genus are widespread in shallow zones of the Southern Ocean. They have been present in the Antarctic since the Pliocene and have survived periods of glacial maximum © Thomas Saucède*

By fragmenting species distribution areas, far from causing their eradication or extinction, the successive glacial-interglacial cycles of the Pliocene in fact contributed to the emergence of new species by favoring repeated episodes of geographical isolation. This phenomenon is known as the *Antarctic Continental Shelf* scenario [PEA 09] or the *Species Diversity Pump* hypothesis [BRA 07]. According to these hypotheses or scenarios, the repeated cycles that fragmented species distribution areas during the advance of ice sheets over the Antarctic continental shelf, followed by the recolonizing of the same areas after the ice retreated, served to stimulate differentiation and the evolution of new species. These hypotheses have been proposed to explain the high species diversity observed among mollusks and echinoderms.

6

Adaptation of Organisms

The environmental conditions prevailing in the Southern Ocean represent extreme values for present-day seas: very low temperatures, permanent or seasonal sea ice cover, darkness during the austral winter, strong currents, strong seasonality, etc. Organisms facing such conditions need very specific coping adaptations. This is all the more true considering the climatic history of the Southern Ocean, which has been marked for over a million years by glacial–interglacial (today's conditions) oscillations that have led to intense natural selection.

6.1. Surviving the cold and escaping the ice

During the austral winter, air temperature drops to below -70°C and winds can be extremely violent, subjecting organisms to extraordinary stresses. Seawater freezes at around -1.9°C and sea ice, from 5 to 7 m in thickness, forms along the coastline, while the ocean becomes covered in sea ice, over an area of almost 20 million square kilometers.

6.1.1. *Fish that make their own antifreeze*

At sea, the first obstacle to overcome is to avoid freezing. This difficulty is revealed by a coastal zone that is generally bare and where few organisms live, which is not the case in seas with more temperate waters. A few meters deeper, in open water, the density and richness of populations can reach impressive levels, but the constraints imposed by ice remain.

For some organisms, such as crustaceans, sea urchins or starfish, the salinity of their internal fluids ensures their survival. Generally, with salinity close to that of seawater, osmoconformity enables a liquid state to be maintained. Furthermore, small crustaceans excrete certain frozen tissues and are potentially capable of surviving temperatures of -5°C in supercooled water [HAW 08]. Mollusks like the limpet *Nacella* secrete slimy mucus that delays freezing [HAW 10]. These limpets are among the rare organisms able to survive in intertidal zones that freeze in winter.

Among vertebrates, mammals and birds are warm-blooded and maintain internal temperatures well above 0°C, by vastly increasing the amount of energy consumed. In contrast, teleosts (bony fish) have a body temperature identical to that of the surrounding environment. They avoid their blood and other internal fluids freezing by secreting antifreeze proteins, controlled by growth hormones, combining their effect with that of salt, also present in small quantities [LEC 04]. These fish could even survive in waters at -2.2°C, if such waters existed. A gene from a pancreatic enzyme is used to produce these proteins. Antifreeze proteins appeared independently in several groups of fish, including five families of notothenioids that were thus able to adapt to Antarctic waters (see Figure 6.1). This is a good example of adaptive convergence. Antifreeze proteins prevent the fish's blood from freezing by bonding to microscopic ice crystals as they form. This has a surprising secondary effect, which was discovered in September 2014: an unexpected property of these proteins being that they prevent the microscopic ice crystals thus formed from reverting back to a liquid state when the temperature rises a little [CZI 14]. Pending further investigation, it is presumed that the advantage brought by this antifreeze function is greater than this particular drawback (preventing complete return to the liquid state).

Figure 6.1. *Crocodile icefish (*Pagetopsis macropterus*)*
© S. Iglesias, CEAMARC

6.1.2. *Looking out for number one, but stronger together*

The strategy adopted by penguins was immortalized in the film *The March of the Penguins*. When winter comes, they crowd together, tightly packed in a Roman army testudo (tortoise) formation, thus conserving considerable heat in the center of the group (temperatures can reach 37°C). A permutation is established, regularly drawing in colder individuals from the periphery, to benefit from the heat in the center. Contrary to appearances, this efficient permutation is not the result of coordinated behavior within the group, but appears to result from the sum of selfish individual behaviors, pushing each individual to move around the group to shelter from the wind, thus displacing the original center, penguin by penguin, towards the side exposed to the wind, and bringing those exposed to the wind into the new center. Mathematical modeling of this dynamic system shows that heat is spread equally among all the penguins, thus optimizing group survival.

6.1.3. *A good insulator*

Eight pinniped species, seals and sea lions, inhabit the Southern Ocean. Some, like the fur seals, do not go south beyond the sub-Antarctic islands or the Peninsula, with its slightly less harsh climate, while others, like the Weddell and crabeater seals, live all year round in the High Antarctic (see Figure 6.2). They are protected against the cold by their short, dense fur, which is more effective against wind, and a thick layer of fat that insulates their vital organs. This layer of blubber can be almost 5 cm thick in Weddell seals, representing 40 to 50% of their body mass in winter. These animals must also adapt their behavior to winter conditions, in particular so that they can continue to feed.

Figure 6.2. *Crabeater seal (*Lobodon carcinophaga*) © Bruno David*

Crabeater seals feed on krill that they dive for. In summer, between two dives, they bask on the ice, enjoying the warmth of the sun. When winter comes, they can travel for hundreds of kilometers to retain access to breathing holes, fractures in pack ice or polynyas (open water pockets), but they always remain close to sea ice or to the coast, in order to rest. In winter, they dive longer and deeper, as food becomes scarcer.

The Weddell seal is the southernmost mammal. It feeds mainly on fish and squid, but also on some mollusks or crustaceans. These seals are able to dig through the ice with their teeth so as to create a passage in and out of the water. Although its thermal conductivity is 25 times higher than that of air, thus significantly increasing heat loss, water occasionally provides refuge for Weddell seals, allowing them to avoid the coldest air temperatures, particularly during blizzards. Apart from that, their behavior is very similar to that of crabeater seals.

6.1.4. *Adaptations in physiology and metabolism*

For 40 million years, the evolution of physiological processes in southern species has kept pace with cooling temperatures, thus maintaining a level of performance compatible with survival, leading to unique adaptations.

The optimum temperature for notothenioid fish is 0°C. It is at this temperature that the ratio is most favorable between energy used and resulting activity. This result is achieved through enzymes that perform best at low temperatures. These "cold" proteins are more flexible, which means that they perform better. The price to pay for this adaptation is that these fish have a thermal tolerance range of under 10°C (between -2°C and +6°C), much smaller than for temperate or tropical fish.

Marine invertebrates of the Southern Ocean lack certain proteins that would allow them to cope with heat shock. Present in most animals, these chaperone proteins control the three-dimensional configuration of other proteins, which thus maintain proper function during temperature changes. Their absence represents significant energy saving for Antarctic marine species living in stable temperature conditions, but restrains their adaptation to variation.

As water temperature decreases, oxygen solubility increases. Better oxygenated water and fluids decrease the cost of oxygen transport to organs, notably muscles. In Antarctic fish, hematocrit (red blood cell count) and capillary density are low. Antarctic notothenioids have two to three times fewer capillaries than sub-Antarctic species [EGG 02]. Hemoglobin diversity also decreases, from four types in teleosts to only two in notothenioids. The blood of channichthyids or icefish is completely devoid of hemoglobin. In these fish, oxygen transport is provided by direct diffusion from the gills to the blood, and then from blood to muscles [LEC 04]. This physiology is associated with anatomical changes – a large heart, large-diameter blood vessels, a greater volume of blood – that allow these fish to reach a good size (70 cm) and to swim actively. The loss of hemoglobin has occurred on four separate occasions during icefish evolution, over the past four to five million years.

Like seals, penguins are protected from the cold by a thick layer of fat, that can represent up to 30% of their body mass at the end of the fattening period that precedes their long winter fast. Such a high proportion of fat would be problematic for humans, but penguins vitally need it and they even manage to avoid using it up too quickly, through mitochondria that allow them to optimize cellular energy efficiency by acting as thermal regulators [FRE 07].

6.2. Living with ice

Surprising as it may at first seem, some species have established their ecological niche in contact with the ice, or near it. This covers very different conditions, however, depending on whether the ice referred to sea ice, no more than a few meters thick, which breaks up into drifting pack ice every summer, or whether it refers to the permanent ice shelves, several hundred meters thick, which are extensions of the polar ice sheet. Dissociated from the substratum, and floating above it, ice shelves cover the Ross Sea and the Weddell Sea, and form huge cliffs jutting out in many places along the coast. Around Antarctica, the ice leaves only 1,700 kilometers of permanently ice-free coastline where intertidal life can develop. Adapting to the presence of ice is therefore essential.

6.2.1. *Sea ice habitats*

Sea ice forms in winter, then melts more or less completely in summer, transforming into pack ice that drifts northward. The lower surface of the ice, its depths and its edges provide an environment that shelters microalgae and bacteria at the base of an entire trophic network Although sea ice appears homogeneous at first, it is actually very varied. The ice is of different ages. The ice floes are mobile and not inter-connected, but able to turn or even fuse together (see Figure 6.3). Salinity varies with depth, from desalted, with snow, at the surface, to brine in places at its base, and ice density also varies (see Chapter 2, section 2.3). Such variety produces, from surface to base, many different micro-habitats, with small salty surface puddles, then cracks containing pockets of brine, denser layers at the base, and platelets beneath the ice [KNO 07].

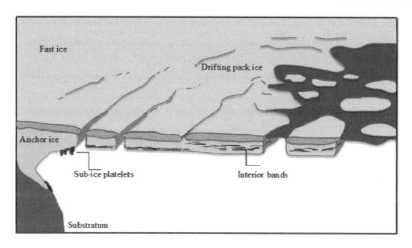

Figure 6.3. *Micro-habitats associated with sea ice. Modified from [KNO 07]*

When air temperatures and sunlight cause melting, small pockets of low salinity water form on the surface. They are colonized by diatoms and flagellates.

In cracks and drains down from the surface, there is a mixture of algae from above with diatoms and bacteria from the base. Salinity can increase in brine pockets formed in these cracks. These pockets are colonized by numerous dinoflagellates, accompanied by ciliates.

Intermediate bands may appear if the ice forms beneath a pre-existing, already colonized surface. Diatoms and dinoflagellates are the most abundant organisms in these intermediate bands.

Diatoms capable of anchoring themselves to ice crystals or of slipping between the crystals form a micro-community in the gaps at the base of the ice, lining its lower surface and encrusting themselves more or less deeply in the ice.

Light can filter through the sea ice, when it is translucent and not too thick, and so microscopic algae (mainly diatoms) can aggregate into greenish-brown filamentous mats, forming ribbons tens of centimeters long.

Disjointed ice platelets also accumulate at the base of the seaice. They form an environment that is 80% water, and thus very favorable to the installation of dense microbial communities, due to the multiplication of available surfaces.

This teeming microscopic life clinging to the underside of the sea ice pack flourishes in lowlight, suggesting a physiology able to survive significant salinity stress, whether hypo- or hypersaline and, for chlorophyll-based organisms, a heightened ability to trap and use light.

Life under the sea ice is not limited only to microscopic colonies of bacteria, algae or small ciliates. Heterotrophic protozoans consume bacteria and algal detritus. Below the ice, the sea teems with larval stages and small-sized adults belonging to various crustacean groups (amphipods, copepods and krill larvae). Some of these animals have developed adaptations that allow them to survive beneath the ice for the entire winter period. Fish larvae, mainly those of the silverfish *Pleurogramma antarcticum*, will in their turn consume copepods. The fish species *Trematomus borchgrevinki* belongs to this cryo-pelagic community. It is a predator, feeding mainly on crustaceans and planktonic gastropods (see Figure 6.4). Much larger, but still belonging to the same vast biological community, are the Minke whales, which are able to circulate through the drifting ice and draw closer to the coastal pack ice, sometimes seeking refuge or even becoming trapped in polynyas (small Minke whales have been observed to spend the winter trapped in polynyas). Their diet is 90% krill and crustaceans found near sea ice.

Figure 6.4. Trematomus borchgrevinki, *a fish from a cryo-pelagic community*
© *Oliver Rey, ICOTA-IPE*

6.2.2. *Far from the world, under the ice shelves*

Hiding deep down beneath the huge thickness of the permanent ice shelves, life subsists in darkness, with no possibility of exchanges with the surface, in very particular conditions. Due to the difficulties of access, life and its adaptations under the ice shelf are still largely unknown, except when something unusual occurs, like the event in March 2002 on Larsen, which released a "plate" with the surface area of Luxembourg, or when expeditions drill through the ice (see Figure 6.5).

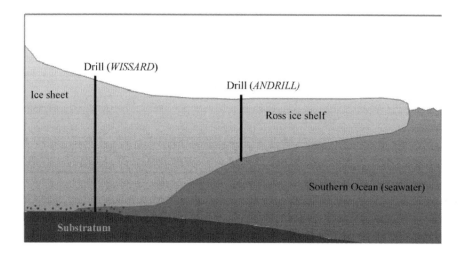

Figure 6.5. *Diagram of the Ross Ice Shelf showing the position of two exploratory boreholes, Andrill and Wissard (figure not to scale)*

One of the most amazing life forms beneath the ice is undoubtedly the small anemone, *Edwardsiella andrillae*, discovered during the ANDRILL expedition in the Ross Sea. Dozens of these small anemones cling to the lower surface of the ice, anchored "head" downward, beneath 250 meters of ice, 40 meters above the seabed [DAL 13a]. How did these anemones adapt their anchoring system to fix on to an icy surface constantly melting and renewing itself? Polychaete worms, amphipod crustaceans and fish that swim upside-down have also been observed in this environment, using the ice shelf above them as their "seabed".

Even more extreme, WISSARD, an American expedition, drilled through the Ross Sea ice shelf in January 2015 to discover one of the most remote ecosystems of the planet, beneath 740 meters of ice [FOX 15]. Rock fragments are constantly falling from within the ice, preventing the development of sessile organisms, such as sponges, so common elsewhere around the continent. There are no traces of life on the seabed, whether epibenthic (living on the seafloor) or endobenthic (buried in the sediment), not even at themicroscopic level. This characteristic is truly extraordinary, because even the most remote abyssal sediments are always populated with microfauna, worms, etc. In this ecosystem beneath the ice, where the depth of the water column does not exceed 10 meters, only mobile organisms have been found: three species of fish, amphipod crustaceans and some other invertebrates. Most surprising of all is that, for organisms to settle here, energy is required. Where can the organic matter required for the development of life here possibly come from? An initial hypothesis suggested by scientists is that it could come from the external surface of the Ross ice shelf, 850 kilometers away, therefore suggesting a journey of 6 to 7 years under the ice, before it would reach the borehole site. Other hypotheses are that energy could come from hydrothermal seeps, or sediment particles released from the ice. Whatever the source, it must be recognized that the animals living here have developed exceptional adaptations, for optimal use of such rare and valuable energy.

6.3. Dealing with intense fluctuations

Due to its polar position, the Southern Ocean is marked by a seasonality that grows stronger approaching the South Pole: with lengthy periods of polar day or polar night, in seasonal alternation with variation in ice cover, and interruption of phytoplankton primary production. In contrast, other parameters, such as the permanently low temperatures, are relatively stable.

6.3.1. *Hellish coastline conditions*

The intertidal zone is the home of seasonal variation and alternations linked to tidal cycles, making it a particularly inhospitable environment. In summer, at low tide, the temperature can rise significantly in the pools of water that remain behind, exposed to the sun, while at the same time melting glaciers bring fresh water, thus decreasing their salinity. In winter, brine circulation can cause local increases in salinity, while decreasing temperatures transform seawater into sea ice. For many years, it was thought that only mobile species, able to migrate towards less exposed areas in winter, could colonize the coastline during the summer. It has recently been shown that several species of bryozoans, crustaceans and mollusks have succeeded in permanently occupying this coastline [WAL 06]. Among these species, the limpet *Nacella concinna* is a flexible migrator. Some individuals, notably the younger ones, remain close to the surface. The protective role of the mucus they secrete (mentioned earlier) allows them to survive supercooling, down to around -10°C and, although mortality increases, some limpets trapped in the ice can survive at temperatures down to -20°C. Animals that take refuge in small water pockets under the ice have a slower metabolic rate, but remain active. They continue to feed and are not in a dormant state where all activity has ceased [OBE 11].

6.3.2. *Advantaged trophic groups*

The absence of phytoplankton production during the austral winter, combined with very low contributions from the frozen continent or from islands with little vegetation, has favored the selection of diets that contribute to the uniqueness of Southern Ocean ecosystems.

Suspension-feeding assemblages (sponges, ascidians, bryozoans, feather stars, and some brittle stars) in particular have developed here. The abundance of rocky fragments torn from the continent and dropped by icebergs favors sessile forms of life. Phoresy, a relationship in which a host organism transports a symbiont or symbionts, is quite common. It is prevalent in large crater sponges that provide shelter to sea cucumbers, brittle stars and crustaceans. It is systematic for sea urchins of the order Cidaroida, as their spines, without epithelium, serve as anchors for many other species. Around fifty different species, mainly sponges, have been counted living on *Ctenocidaris spinosa* (see Figure 6.6).

Figure 6.6. *The sea urchin* Ctenocidaris rugosa *(around 15 cm in size) with spines covered in symbionts. Sponges are visible on the left and in the center, with a sea cucumber on the right © Biogéosciences*

To cope with such intense fluctuations, organisms have the "choice" between spending the winter surviving on accumulated reserves and fasting, or adopting an opportunistic diet. Many sessile suspension feeders adopt a slower lifestyle. Krill (*Euphasia superba*) accumulate greater lipid reserves the further south they live [SCH 14], and their metabolism slows down, optimizing use of their reserves. Other mobile species adapt their diet to the resources available. The more southern sea urchins are omnivores, while their relatives in sub-Antarctic and Magellanic regions are either herbivores or carnivores. Among fish, the diet of Antarctic notothenioids is more varied than that of sub-Antarctic species. The starfish *Odontaster validus*, the sea urchin *Sterechinus neumayeri* and other benthic invertebrates of Adelie Land become scavengers in winter. In fact, *O. validus* has been described as "a rather unselective omnivore" [PEA 69], feeding on anything it can find, from diatoms and small ciliates to sea urchins, carrion and Weddell seal feces.

6.3.3. Feeding their young by endless periods of fasting

Emperor penguins (*Aptenodytes forsteri*) reproduce in the middle of the austral winter. The female lays a single egg that is incubated by the male while she travels hundreds of kilometers to hunt for food. After the egg has hatched, its parents take turns to hunt for food on the sea ice and in the sea, which they regurgitate weeks later to feed the chick. If the chick hatches before the mother returns, the male, deprived of food for over 100 days, feeds the chick with a secretion produced by an esophageal gland. King penguins (*Aptenodytes patagonicus*) of the sub-Antarctic islands face the same problem, amplified by the recent slight warming of the waters in that region, thus requiring longer journeys for the female to find food resources. While waiting for the female, the male feeds the chick with undigested food stored in his stomach. This bolus is protected by an antimicrobial peptide of the defensin family, spheniscine, which prevents the development of pathogenic bacteria and fungi, and which increases in concentration during food storage periods [THO 03].

6.3.4. From total night to permanent day

A unique characteristic of polar seas is the speed of the shift from almost complete winter darkness to permanent summer daylight. From one week to the next, daylight increases or decreases by one hour. The organisms that depend on daylight to move, feed or respire have had to adapt to a rapidly changing photoperiod, which must have meant developing mechanisms that can quickly modulate their vital functions. Experiments on krill have shown that these crustaceans express more genes involved in metabolism and motor activity during periods of daylight. The proteins necessary for molting have proven to be especially sensitive to photoperiod [SEE 09]. The impacts of such changes in light availability are even more marked in photosynthetic organisms such as microalgae and phytoplankton.

6.4. Lower metabolic rates, longer lifespans and gigantism

Cold, ice, seasonality of resources, and the intensification of these conditions, which reaches its peak during glacial episodes, have favored the emergence and selection of characteristics that are neither imperative nor exclusive to the Southern Ocean, but that are nevertheless more common here than in other oceans, including the Arctic Ocean. Life history traits (e.g. growth and longevity) are specifically targeted by these adaptations. Many ectothermic species from Antarctic marine environments, subjected to permanent frost, tend to reduce their metabolic cost and slow down their activity in a way that enables them to maintain growth, even by extending their development period if necessary [PÖR 06].

6.4.1. *Metabolism and development*

The metabolism of species that live in the Southern Ocean can be very different, depending on whether they are endotherms or ectotherms. Endothermic mammals and birds, which need to maintain high internal temperatures, must have a very active metabolism (aerobic when out of the water, anaerobic when diving). In contrast, ectothermic fish and invertebrates have a slower metabolism, with reduced energy costs. Thus, in various mollusks and anemones, locomotor activity and burying behavior are reduced or slowed down by half, compared to observations for similar temperate species [PEC 06]. The same applies to digestion, which is two to ten times longer in comparison with temperate species. The low metabolic rate of many Antarctic marine species mitigates the drawbacks linked to the reduction in primary productivity for several months a year (only 2 to 3 months in the High Antarctic). These negative effects of temperature on metabolism are partly compensated by the availability of dissolved oxygen, which improves efficiency compared to temperate forms, but the limiting factor remains access to food.

Some animals opt for a form of quasi-hibernation. Observations of *Sterechinus neumayeri* (see Figure 6.7) on Adelaide Island show that these sea urchins are able to stop eating for 7 months without intensive use of reserves [BRO 01a]. Slowing down their metabolism is the basis of this adaptive strategy. Their oxygen consumption, which is already low in summer, is divided by three in winter [BRO 01b]. The situation is more complex for fish swimming activity. Fast swimming is strongly affected (it uses white muscle, which draws energy from anaerobic glycolysis), whereas suspension swimming is maintained at a stable level, due to an increase in the number of mitochondria in red muscle cells, a physiological compensation for the effects of cold, with increased energy cost [PEC 02]. The decline in functional efficiency of each mitochondrion is compensated by their number. Resource allocation thus focuses on the most essential type of swimming. In contrast, in the limpet *Nacella concinna*, muscle efficiency is lower and ability to anchor to a substrate has been measured as 2 to 20 times less than for temperate limpets [DAV 88].

Figure 6.7. *The sea urchin* Sterechinus neumayeri *(size around 5 cm), in the center (with an algal "hat") and at the bottom. King George Island, depth 7 m © Bruno David*

The Q_{10} index expresses the rate of change in a biological system subjected to a temperature variation of 10 °C. In general, Q_{10} values recorded in marine invertebrates of the Southern Ocean are low as long as temperatures remain cold, but the index increases rapidly once the temperature exceeds the natural thermal range, indicating strict adaptation (stenothermia). This index is considerably higher than in equivalent temperate forms, indicating great metabolic sensitivity to temperature increase in Antarctic organisms, and highlighting the limits of their adaptive capacity. In Antarctic echinoderms, when temperature increases, the Q_{10} index observed for their development is 3 to 7 times higher than predicted by thermodynamics, showing that factors other than temperature are also involved: seasonality of resources and of light, factors perhaps inherited from history and from colder periods.

Development is also slower. With few exceptions, growth rates for Antarctic benthic species are two to five times lower than those for temperate species of the same groups [PEC 06]. As a result, the first stages of development (embryonic phase) are often very long. The time lag can be very lengthy: from 3 to 15 times slower in Antarctic mollusks than in temperate species, and from 6 to over 50 times slower in brachiopods. The groups most affected by these extended embryonic phases are brooding crustaceans and gastropods. The isopod *Glyptonotus antarcticus* incubates its embryos for 20 months, instead of 1 to 3 months for temperate forms. The record is held by the gastropod *Neobuccinum eatoni*, which incubates for 25 months [PEC 02]. This slower development results in later acquisition of sexual maturity and an extended lifespan.

6.4.2. *Long-lived forms*

In correlation with slow growth rates, which are linked to a slower or discontinuous metabolic rate (with an arrest phase in winter), organisms in the Southern Ocean often have long to very long lifespans.

Assessing the lifespan of marine species living in the depths is still rarely attempted and so there is significant uncertainty surrounding the estimates that can be made, especially for species other than the most common ones, notably those that are of commercial interest, for which it is necessary to set up quotas and safeguards. This limitation is even truer in the Southern Ocean, where difficult access makes temporal investigations and tagging much more problematic. Despite these practical problems, several specific cases have been explored in detail, and indirect age indicators have also been used.

Initially, it seems simple to say how old an organism is, as we are so accustomed to knowing our own age, or that of the living beings that surround us (calves, cows, pigs, or trees). In reality, in the absence of the temporal monitoring of development that exists on farms, for example, the age of an organism is only accessible by indirect approaches, with a greater or lesser margin of error.

The most favorable case is where growth is regularly recorded on the organism itself. Successive phases of rapid or slow growth produce visible or easily revealed growth lines on the shell, or the skeleton, or more discrete elements, such as otoliths (earstones) in fish. The hypothesis is that these growth lines are the expression of a regular and stable phenomenon over time, such as day and night or seasonal alternation. By simply counting growth lines, age can then be calculated.

The worst case scenario is when there is no record of growth to permit calibration. In this situation, a population-based approach is used. Starting with a sample of individuals of all ages and sizes, it is then necessary to distinguish between general group variation (for the same stage of development, not all individuals are identical) and what stems from development itself (young specimens are different from adults). The growth of a virtual individual is then plotted on to the morphological evolution of a population, with each member of the population representing the virtual individual at a given time in its life. Size and distribution patterns can be used, together with collection logs, to detect cohorts, providing a rough idea of age.

After these few cautionary remarks, the following observations remain valid. Antarctic sponges are believed to be centuries old, with one of them (*Cinachyra antarctica*) given an age of 1,550 years. The starfish *Odontaster validus* has been declared a centenarian, based on aquarium rearing (see Figure 6.8). Other organisms also live for a relatively long time, without reaching record age for their groups, e.g. the fish species *Trematomus* (more than 20 years) and *Dissostichus* (39 years), or the sea urchins *Sterechinus neumayeri* and *S. antarticus* (more than 50 years), and the brachiopod *Liothyrella uva* (55 years) (see Figure 7.2, Chapter 7).

Figure 6.8. *The starfish* Odontaster validus *(size around 15 cm). King George Island, depth 5 m © Bruno David*

With an estimated growth of 1 mm per year, the endemic Antarctic solitary coral *Flabellum impensum* could take 80 years to reach a size of eight centimeters, which would be very large for this group [HEN 13]. On Signy Island (South Orkney Islands), the growth rate of the octocoral (sea fan) *Primnoella scotiae* is five times slower than that observed for most other cold-water octocorals from other regions, and at least 1.7 times slower than the slowest rate known elsewhere [PEC 13].

6.4.3. *Gigantism*

For divers accustomed to seeing benthic fauna in the temperate or even cool waters of European seas, brittle stars (echinoderms with five thin arms, relatives of starfish) are small, slender animals that can be very abundant locally, for example on the sandy seabed off the coast of Normandy or in the south of England. The most common brittle stars in the Channel or the Mediterranean Sea (*Ophiura ophiura* and *Ophiotrix fragilis*) reach adulthood at 7 to 10 cm. Another common species (*Amphipholis squamata*) is somewhat smaller (2–3 cm). In the tropics, many species that live clinging to branches of coral are also very small and considered to be dwarfs. Diving along the Antarctic coast, scientists have found brittle stars, like *Ophionotus victoriae*, which exceed 20 cm in size. *Ophiosparte gigas* is, as its name suggests, a giant brittle star measuring more than 40 cm in diameter, with a 7 cm disk. Some starfish could be described as "monsters", e.g. representatives of the *Labidiaster annulatus* species, which have over 40 arms, or *Macroptychaster accrescens*, which exceeds 60 cm in diameter and weighs more than 5 kilograms.

Bonellians are echiuran worms: the female has a T-shaped external proboscis, which can be extended to catch food particles or small organisms and transport them to its mouth. The common species found in the Mediterranean, *Bonellia viridis*, has a proboscis that can reach up to one meter when extended. In Antarctic waters, related species can reach a size of over 3 m.

With long or even very long lifespans, and with late onset of sexual maturity accompanied by continued growth, some Antarctic species reach relatively large sizes. Furthermore, increased oxygen solubility in cold water is associated with better diffusion capacity, making a larger size possible without too significant an increase in energy cost. The record for size in sponges is held by a hexactinellid (*Anoxycalyx jouboni*) from the Ross Sea, measuring 2 m in height, with a diameter of 1.50 m [DAY 71]. The ribbon worm (nemertean) *Parborlasia corrugatus* can reach up to 2 m in length, rather than a more usual size of several centimeters [DAV 02]. The pycnogonids, arthropod cousins to spiders, and usually very discreet (infra-centimetric size), are a group where giant southern forms are common, and can reach around fifteen centimeters. This frequent gigantism also affects amphipod and isopod crustaceans, as well as ascidians (see Figure 6.9). Other groups, such as sea urchins, gastropods or sea fans, however, remain at a size comparable to that of their cousins in other regions [ARN 74a]. As is the case in other seas around the world, there are also dwarf versions of some forms, for example, calcareous sponges, calcareous foraminifera, brachiopods, some gastropods, etc.

Figure 6.9. *The amphipod* Eusirus properdentatus. *Specimen from the Weddell Sea measuring almost 10 cm, while the usual size of representatives of the group is around one centimeter © Thomas Saucède*

The frequency of these forms of such great size is explained by the stability of the environment, which allows long-term "investments" and has thus permitted, over the course of evolution, the selection of very long-lived forms, such as those observed in the abyssal depths. Other characteristics have also contributed to the selection of giant forms, and their origin is probably multifactorial [MOR 12]. Possible factors are oxygen availability associated with low metabolic rates, while water chemistry particularly rich in silicates and poor in carbonates could explain dwarfism in organisms with calcareous skeletons. Deep ocean species are also adapted to cold waters and to special, but stable, environmental conditions. Quite a few cases of gigantism have been reported in the abyssal world and one hypothesis is that part of Antarctic fauna could be of deep origin (the polar emergence hypothesis described in Chapter 5, section 5.2). It is nevertheless awkward to propose just one scenario to explain the frequency of giant forms in the Southern Ocean. It is a complex issue, since historical and environmental factors are involved, as well as species life history traits.

6.5. Parents caring for their offspring

Taking care of one's own children is natural … at least for *Homo sapiens* in the 21st Century. It seems even more obvious with daily observations of this behavior in many familiar animals, beginning with cats, dogs, and horses, but also in blackbirds, blue tits, or robins in our gardens and forests. However, many species are content to bring a large number of descendants into the world, and let circumstances decide which ones will rapidly disappear (the vast majority), and which will survive to reach sexual maturity and reproductive age, allowing a new generation to be born.

6.5.1. *Two strategies*

MacArthur and Wilson mapped out the adaptive strategies of organisms by opposing two options designated by the letters r and K. The r parameter refers to the growth rate of a population (the "intrinsic rate of increase"), while the K parameter refers to the "carrying capacity of the environment". The higher the growth rate of a population (each individual has a large number of descendants), the more rapidly the carrying capacity of the environment is reached (maximum number of individuals that can be present at time t). These two parameters are somewhat antagonistic: either reproduce rapidly or optimize resource use. The general idea is that if resources fluctuate and are unpredictable, it is in the best interest of a species to reproduce as fast as possible, in order to benefit from favorable conditions when they occur, and to exploit resources to the maximum before they disappear. Conversely, in an environment that is stable or predictable (with regular fluctuations), it is in the best interest of a species to optimize the exploitation of resources, even by reducing reproduction. This is what led MacArthur and Wilson to define two strategies that are the extremes of a continuum. The r strategists have very high fecundity and mortality rates. They are often small-sized species that reach sexual maturity very early and have very short generation times. Parents do not take care, or take little care, of their multiple offspring, many of which die rapidly. Many micro-organisms are r strategists. In contrast, K strategists are long-living species, reaching sexual maturity quite late and producing few descendants. However, they take care of their offspring, thus maximizing their survival rate. Cetaceans are emblematic of this strategy. The r strategists have "productivist" logic and invest in the quantity of their descendants. The K strategists have a logic of "quality" and invest in the survival of adults.

There is therefore a link between adaptive strategy, longevity, maximum size reached and parental care for juveniles. In the sea, species with external fertilization and larval stages are generally r strategists. This is the case for many benthic invertebrates, except in the Southern Ocean.

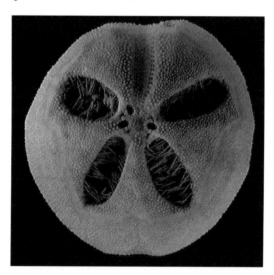

Figure 6.10. *A female of the irregular brooding sea urchin* Abatus cordatus, *Kerguelen (size around 3 cm). The specimen has had its spines removed. The three small orifices in the center are gonopores. The four depressions are brood pouches © Biogéosciences*

6.5.2. *Kangaroo sea urchins*

Sea urchins are dioecious (males and females are separate individuals). When it is time to reproduce, both males and females release millions of gametes and fertilization occurs in open water. The fertilized egg grows a little larger, and then produces a small larva (the pluteus), which drifts with the currents, feeding on phytoplankton for several weeks, before moving closer to the seabed and metamorphosing into a young sea urchin. This strategy allows dispersal over vast distances, ensuring the colonization of distant sites and the perpetuation of a certain homogeneity of the species through genetic mixing. The trade-off is that many larvae will be lost in environments unsuitable for their survival or will become victims to predators. Some species, notably in the deeper domain, produce larger eggs, rich in nutrient reserves. The resulting larvae float and swim very little, and instead remain close to the bottom, consuming egg nutrient reserves to survive, and thus limiting dispersal and its associated risks. An additional stage exists in many southern species, all called kangaroo sea urchins, although they belong to different groups.

Kangaroo sea urchins are species where the female produces very large eggs, very rich in nutrient reserves. Once fertilized, these eggs develop directly into young sea urchins without passing through a larval stage. The resources necessary to produce such eggs are considerably greater than for normal-sized eggs, so the females take great care of them. They shelter them in brood pouches (marsupia), where the first stages of development will take place. In different groups, marsupia may have very different anatomical structures, confirming that this adaptive strategy has developed independently at least three times in Southern Ocean sea urchins. Regular cidaroid sea urchins incubate their young in shallow depressions surrounding the mouth. The brooding species of the schizasterid family (irregular sea urchins) have depressed ambulacra, forming four deep pouches on the "back" of the female, with a protective barrier of spines (see Figure 6.10). Although fertilization by sperm is external, it occurs during egg-laying while the eggs are still on the female. Using her spines, the female then pushes the fertilized eggs into the pouches. A female *Abatus cordatus*, a Kerguelen sea urchin, hosts about 200 young for 9 months in her brood pouches. At the end of this time, when juveniles have reached the long spine stage and are able to move, they leave the pouches and begin living free, near their mother.

Another even more extreme case of K strategy in sea urchins was discovered in the early 1990s [DAV 90]. In two species of the genus *Antrechinus* (Southern Ocean irregular sea urchins), the young are incubated in deep invaginations suspended within the body of the female. Each of these brood pouches communicates with the exterior through a birth canal leading to a narrow orifice protected by plates and spines (see Figure 6.11). The oviducts of the female open into the birth canal, suggesting that fertilization probably occurs inside the pouch. Although incubation is not physiologically internal (intra-coelomic), the young are sheltered within the body of the female. The pouches contain at most a dozen juveniles, each reaching 6 to 7 mm in length, with long spines of around 5 mm, which is relatively large when compared to a mother measuring around 30 mm in size. To allow the young to move outside, the birth canal dilates and the plates surrounding the orifice open by means of collagen hinges. The emergence of the young is therefore akin to a form of birth, even though it is not viviparity.

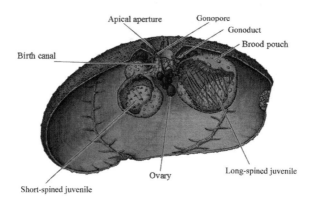

Figure 6.11. *Female of the brooding sea urchin* Antrechinus mortenseni © *Rich Mooi*

This anatomical organization and associated reproductive processes are unique to sea urchins, and represent an extreme case of *K* strategy and protection provided to eggs and juveniles. Around 80 species of sea urchins live in the Southern Ocean and among them more than 50% are brooders (from 42 to 45 species). In the rest of the World Ocean, including the Arctic Ocean, there are only 9 brooding species, which is only about 1% of existing species. What a contrast!

6.5.3. *Why is there so much brooding in the Southern Ocean?*

The frequency of *K* strategy and the brooding that accompanies it is not restricted only to sea urchins, and nearly all groups have species that adopt strategy. In demosponges (siliceous sponges), more than 80% of southern species are brooders, whereas the proportion is only about 50% in other seas, except the Arctic Ocean, where there is a similar percentage to that of the Southern Ocean [ARN 74b]. In the solitary coral, *Flabellum*, the three Antarctic species are brooders, whereas other species of the genus living in the cold North Atlantic waters are broadcast spawners that produce lecithotrophic eggs. As a corollary, Antarctic species produce oocytes about five times larger than those of other species [WAL

08]. Numerous bivalve species protect their young in their mantle cavity or between the gill lamellae. Some are very marked K strategists, with the number of descendants reduced to a few dozen eggs or embryos. Earlier in this chapter (section 6.4.1), the incredibly long incubation periods observed in the isopod *Glyptonotus antarcticus* and the gastropod *Neobuccinum eatoni* were mentioned. A very similar situation to that observed in sea urchins is found in other echinoderms, like starfish and brittle stars: around 50% are brooders, a much higher percentage than that observed in other seas, including the Arctic. Many other examples could also be given.

What factors might account for this phenomenon? The contrast often observed with the Arctic Ocean, where many groups have few brooding forms, leads to the rejection of a hypothesis based on temperature effect and harsh polar conditions. It must, however, be stressed that although brooding is much less common in the north than in the south, the K strategy is still adopted by many species. Although they do not brood their young, these species produce large eggs, in smaller numbers, that become larvae feeding on egg nutrient reserves and remaining close to the seabed (demersal larvae). A hypothesis applicable to both the polar seas could be based on their contrasting seasonality and accompanying ice cover, a situation leading to brief and not always synchronous plankton cycles from one year to the next. This relative unpredictability could have led to the selection of direct development in the south (brooding) or shorter development in the north (demersal larvae), thus overcoming the reliance on plankton cycles. Paradoxically, therefore, this resource instability leads to a K strategy. However, this hypothesis does not take into account the concomitant existence and development of pelagic larval forms.

Why is there such a difference between north and south? The explanation could perhaps lie in the history of the groups and the specific context of the Antarctic and its biodiversity. The isolation of the Antarctic continent is the result of a long history that began with the break-up of Gondwana (see Chapter 3, section 3.1). The last stage, the opening of the Drake Passage, marking the separation from South America and the establishment of the Antarctic Circumpolar Current (ACC), dates to the mid-Eocene, around 40 Ma. This period was followed by intensification of cooling, around 34 Ma. This means that the fauna and flora of the Southern Ocean have been isolated, spatially and ecologically, for a very long time, much longer and in a much stronger way than those of the Arctic Ocean. Another important historical factor is that, for many groups, Antarctic biodiversity shares common ancestors with groups of other regions in the Southern Hemisphere, notably Australia. Yet many of the ancestors that developed in warmer waters were brooding species. The evolution of Southern Ocean fauna was thus rooted in groups bearing this characteristic, which is not the case in the Arctic. This scenario is well illustrated by sea urchins, where only one brooding species has persisted in Australia, while there had been 10 such species in the Cenozoic Era, before this island-continent separated from Antarctica

[PHI 71]. In the north, although brooding sea urchins have been identified from the Pliocene in western France, they left only a single descendent now living in the deep waters of the Atlantic [DUD 05].

Another phenomenon has contributed to the success of brooding forms in the Southern Ocean. Over the last 2.5 million years, and especially in the past 800,00 years, there have been many successive glacial-interglacial cycles. During glacial episodes, the extension of permanent ice prevented or severely restricted the production of plankton over large areas, which would have been an advantage for species with direct development and, as a corollary, selected against plankton-eating species [POU 96]. More probably, the presence of ice led to the isolation of small regions favorable to benthic lifestyles, all around the Antarctic continent. This in turn would favor habitat fragmentation and the isolation of populations in refuge areas. As brooding species do not disperse (the N+1 generation stays close to the N generation), fragmentation and isolation probably induced many speciation events, prompting the increase in brooding species and the radiations observed in some groups, such as the irregular sea urchins.

An alternative but not contradictory explanation involves some individuals being transported toward new sectors by the ACC. The presence of a greater proportion of brooding species in the Scotia Arc than close to the continent suggests that the current may have played this role at the entrance to the Drake Passage, where it is the strongest. In each new sector colonized, the low genetic diversity of the individuals transported would rapidly have favored a drift toward a new species [PEA 09].

Nevertheless, the frequency of brooding species in the waters of the Southern Ocean is more probably linked to speciation processes induced by the polar setting, in particular its climatic and oceanographic history, than to narrow adaptation to specific environmental conditions. The ability to develop a K strategy with brooding, which existed prior to the isolation of the Antarctic, was maintained in different groups that were then subjected to a more intense phase of radiation (species proliferation).

7

Projections into the Future

Projections into the future are a classic way to end a book. They remain a delicate exercise with results that often turn out to be inaccurate in the long term, or even completely wrong. Their only advantage is that, due to their long-term perspective, their authors and readers will no longer be there to check their validity. In fact, such projections can be set for different time scales, all distant. In our lives, everyday situations generally lead us to anticipate several weeks or months into the future (next summer's holidays, the start of the school year, etc.). We are also used to anticipating several years or decades in the future (buying a house, preparing for retirement, and of course the prospect of our own demise). However, such projections, whatever their time scale, are about as rapid as the beating of a butterfly's wings when compared with the future of an ocean and its biodiversity. Human beings are certainly capable, through their actions, of considerably accelerating natural phenomena, but system inertia will absorb the effects of most changes, even by preventing any steps backward. Foreseeing the future of the Southern Ocean is an exercise that could cover centuries, millennia or millions of years, depending on whether the focus is on the impact of human activity or the evolution of major planetary cycles.

7.1. The immediate future

The immediate future means towards the end of this century or the beginning of the next. The most important changes that are expected to occur will probably involve the reaction of natural systems to human activity.

For the past few decades, regular data recordings have become available for the Antarctic. Climate change due to greenhouse gases is marked by a certain stability in the temperature data recorded that masks great regional disparities. The Antarctic Peninsula, vast zones of West Antarctica and the sub-Antarctic areas have experienced significant warming (+3.2°C increase on average in the Peninsula over 60 years, five times the global rate), while other areas appear to have cooled (the center of the continent and East Antarctica). For several decades, this disruption of the climate has been shown by a reduction in the ice sheet on the Antarctic continent, and ice loss (from ice slipping into the sea) has increased from 30 to 140 Gt/year between 1992 and 2011 [GIE 13]. This phenomenon is particularly prominent around the Antarctic Peninsula, and the Amundsen Sea. In the ocean, the effect of this ice loss translates into the much stronger impact of large tabular icebergs that scrape the seabed, destroying coastal benthic fauna down to depths of 200 to 300 meters [GUT 01]. The surface temperature of the Southern Ocean has increased by around 1°C around the Peninsula and the Antarctic Circumpolar Current (ACC) waters are warming quite rapidly, faster than those of the World Ocean. Paradoxically, sea ice appears to have increased (see Figure 7.1). The cumulative increase over the last 30 years is equivalent to the surface area of France, around 500,000 km^2, but with regional disparities [GIE 13].

Figure 7.1. *Sea ice in the Erebus and Terror Gulf, at the entrance to the Weddell Sea, facing Paulet Island. At this time of year (February) the sea should be free from ice © Bruno David*

If projected on to the "immediate future" of the last 20 years in this century, models predict the general warming of all Antarctic regions [BRA 08]. Such warming would be particularly apparent during autumn and the austral winter and would be more marked in the Weddell Sea, in the region of Queen Maud Land and in the Ross Sea. In some areas, it could exceed 5°C. Again, according to these models, warming would be accompanied by a decrease in sea ice, especially in winter and in spring, and by an increase in wind speed, especially in autumn, in the latitudinal band corresponding to the ACC, which might then accelerate.

These climatic phenomena, whether observed or predicted, give rise to various processes affecting not only the ocean (ice cover, ACC and acidification), but also the living world.

7.1.1. *Invasive species*

If protection of the continent and the Southern Ocean is no longer ensured by the extension of the Antarctic Treaty, initiated in 1959 in Washington, it would be safe to bet that economic appetites will take precedence over some sense of the common good of humanity. The first invasive species therefore risks being *Homo sapiens*, almost immediately followed by their commensals, rats, mice, etc., which will adapt all the more easily if significant warming occurs, increasing the surface area free from ice.

At sea, increased shipping coupled with warming water may contribute to the installation of invasive species from cool temperate zones. Already today, the Polar Front and the ACC are no longer impermeable barriers and some organisms can cross them. Such "travelers" are still infrequent and it is rare for them to be successful in adapting to zones where conditions are quite different from those prevailing in their region of origin.

The invaders that may have the strongest impact on Antarctic marine ecosystems are the decapods. Today, this group of crustaceans (crabs, lobsters, etc.) is represented by only 22 species in the Southern Ocean. Most are absent from the Antarctic zone (only 11 survive beyond 60° S) and none have been found on the continental shelf or in the waters of the High Antarctic, where the temperature is below zero [DEB 14]. This limitation is due to their inability to regulate the level of magnesium in their blood, which paralyses and kills them when the temperature drops too low [FRE 00]. Some cold temperate decapod species nevertheless have the ability to keep their magnesium level low enough to become strong candidates for invasion if the water temperature becomes slightly warmer.

The fact that lithodid crabs (*Neolithodes yaldwyni*) have been observed in Antarctic waters has led to a dramatic invasion scenario being put forward, as the arrival of such predators, absent from the region for 15 million years, could drastically affect ecosystems [THA 05]. However, with regard to these crabs, such a catastrophic scenario might not be entirely accurate, since the crabs remain confined to the slopes of the shelf edge, where they may have persisted over a very long time [GRI 13]. For these king crabs, the most likely scenario is the emergence of deep-sea fauna moving toward the continental shelf. Meanwhile, another representative of the same group, the spider crab *Hyas araneus*, native to the North Atlantic, has been observed near the Peninsula, probably transported there in larval form by a ship [TAV 04]. If Antarctic waters become busier and if the water temperature increases a little, it is inevitable that other species will follow and settle there. This is all the more plausible as large anomuran crabs present adaptations (K strategy, slow growth, brooding, etc.) that could be the key to success in future expansion.

Recently, a female brachyuran crab, *Halicarcinus planatus*, common in Patagonia, was found on Deception Island in the South Shetland archipelago [ARO 15]. Although one individual does not constitute an invasion, this volcanic island is the site of intense hydrothermal activity, warming the waters of its caldera. Furthermore, it has become the destination of ever more numerous tourist cruises, which are potential sources of larvae or young individuals transported in ballast water or on ship hulls. These characteristics make Deception Island a potential "Trojan Horse" for future invasions, all the more so since the Peninsula area is the area of the Southern Ocean that is experiencing the most rapid warming. The European green crab, *Carcinus maenas*, an efficient predator dispersed by man on all inhabited continents in the Southern Hemisphere, is a potential invader.

Elasmobranch fish (sharks and rays) are very rare in the Southern Ocean. Sharks are only present in the sub-Antarctic Zone (Kerguelen, Crozet, and South Georgia) or in deep waters. Some species of rays have settled in coastal waters around the continent [DEB 14]. Pelagic sharks swim fast to ventilate their gills, thus requiring a high metabolic rate incompatible with cold Antarctic waters. Even bottom (benthic) sharks, which are relatively sluggish swimmers, seem to face a physiological barrier [ARO 07]. Many sharks live in the cold temperate regions of South America, Australia and New Zealand. If the temperature rises sufficiently, pelagic sharks will easily be able to cross the seas to settle in the Southern Ocean. This will be more complicated for benthic sharks, as they would be hindered by depths incompatible with their survival.

The Southern Ocean benthic fauna is characterized, particularly around the continent, by the absence of apex predators, which are very mobile. This allows easily accessible sessile filter feeders to proliferate there, along with their not very mobile predators, i.e. starfish or nemerteans (worms). This fauna is functionally similar to that of the deep ocean, with something of the Paleozoic about it [ARO 07]. The arrival of decapods, fast predators able to break shells, in ecosystems from which they were previously absent, could have a huge impact on mollusk, echinoderm and arthropod fauna. Some giant, relatively slow and static benthic forms, like the pycnogonids, are not yet facing such predation pressure, which could fundamentally modify trophic networks, leading to new forms of equilibrium that would make Antarctic ecosystems functionally resemble those of other oceans.

Although not an invasive phenomenon in itself, in the case of warming, the decrease in ice cover, with its accompanying erosive effects on the coastline, could favor the extension of the distribution areas for various algal species, which would also benefit from improved light conditions.

7.1.2. Extinctions

When environmental conditions change and become less favorable, the first reaction of affected populations is to acclimatize, for example by adapting their phenology. If waters warm up a little earlier at the end of winter, phytoplankton will bloom earlier, and dependent species will tend to modify their life cycle. A second possibility is to move in order to follow the shifting climate belts. Seeking a new habitat is a quick and easy solution for mobile organisms, but it is more hazardous if the move will take several generations, with the dispersal of gametes or pelagic larvae. In the Southern Ocean, the K strategies selected millions of years ago, in the aftermath of isolation and cooling, are particularly ill adapted to such a process. A slow growth rate lengthens generations and therefore the time required for such a move, and the small number of juveniles diminishes the chances of success, while brooding phenomena reduce dispersal (see Chapter 6, section 6.5). The most important problem facing species endemic to High Antarctica and adapted to very low temperatures, if the coldest waters in the World Ocean become warmer, is that it is impossible for them to move further south. They are caught in a climate "trap" that prevents them from seeking new habitats. A further difficulty is that they risk having to cope with the arrival of climatic migrants from the north. They will therefore have to adapt or disappear, with the latter alternative seeming more probable.

Many ectotherm species in the Southern Ocean have developed coping mechanisms for very low temperatures. In the case of warming, these physiological adaptations could prove to be insurmountable handicaps, especially since some species are stenotherms, with a narrow thermal optimum. This is the case for the brachiopod *Liothyrella uva* (see Figure 7.2), which has an upper lethal temperature between 3.0°C and 4.5°C, or the bivalve *Limopsis marionensis*, which cannot survive in the long term above 2°C [PÖR 99]. Even if experiments in aquariums under controlled conditions have shown that many other species are capable of withstanding relatively high temperatures, these results do not properly take into account two important parameters. The first is the length of exposure to heat stress. A temperature that is non-lethal over a period of a few hours (sunlight, or tidal oscillations) or several days (changing weather) can become lethal over longer time periods. The second parameter is that long-term survival for individuals could still result in a decrease in reproductive success, due to greater allocation of resources for immediate survival, at the expense of reproduction. This process will ultimately lead to the decline and disappearance of populations, and then the extinction of the species.

Figure 7.2. *Brachiopod (*Liothyrella uva*) at a depth of around 7 m. King George Island in the South Shetlands © Bruno David*

Even endotherms may be affected. The warm oscillations of the El Niño phenomenon have an indirect effect on sub-Antarctic populations of king penguins. Both their reproductive success and adult survival tend to decrease. More directly, the probability of survival for a king penguin is an inverse function of sea surface temperature. A small increase of 0.3°C in temperature results in a 10% decrease in survival [LEB 07]. This relationship is not linear and it is fairly safe to predict that, if water temperatures rise, the decline will accelerate rapidly until it reaches a threshold that will abruptly kill off the few surviving king penguins in the affected areas.

For organisms whose survival depends on the presence of ice shelves, the reduction in ice-shelf extent has begun to have local effects. Some populations of Adelie penguins on the Antarctic Peninsula (see Figure 7.3) have declined by nearly 65% and have only partly been replaced by chinstrap and gentoo penguins [DUC 07]. This case illustrates a situation that could become the norm: deleterious selective pressures on a species, to which is added competition from newcomers.

Figure 7.3. *Chinstrap penguins (*Pygoscelis antarcticus*), a species gradually replacing Adelie penguins (*Pygoscelis adeliae*). King George Island in the South Shetlands © Bruno David*

These well-documented examples show that even episodic or small-scale fluctuations in environmental conditions can have a significant impact. For some species, the consequences of climate change are already present. If warming continues, local extinctions are to be expected, accompanied by southward migration wherever possible, with permanent extinction when migration is impossible or too difficult. The remarkable ecosystems of the Southern Ocean could be very much affected by such changes.

7.1.3. *Acidification*

For a little over 20 million years, the World Ocean has had a very stable, slightly alkaline pH, varying between 8.1 and 8.3 [PEA 00]. These oscillations have not been cyclical and the transition has always been slow and gradual, allowing sufficient time for life to adapt.

Between 25 and 30% of the CO_2 emitted into the atmosphere by human activity dissolves into the ocean. The increase in the amount of dissolved CO_2 lowers the pH of the water and thus increases relative acidification; relative because seawater remains alkaline and will never become acidic. Since the beginning of the industrial age, the ocean's pH has fallen from 8.2 to just under 8.1. Most scenarios predict a fall of more than 0.3 units in the next 100 years. This might seem a very minor decrease, but pH is on a logarithmic scale so 0.3 units will mean that acidity will double in comparison with current levels. In the case of the Southern Ocean, there is an additional parameter to consider. Indeed, acidification kinetics is not identical for all the oceans in the world. It is faster when the waters are colder, because their dissolving capacity is greater. This means that the cold Antarctic will be affected much more quickly by a fall in pH, while awaiting the homogenization that will occur in the much longer term, through thermohaline circulation.

The consequences of this fall in pH could be very significant or even dramatic for calcifying organisms (foraminifera, corals, mollusks, bryozoans, crustaceans, etc.), which are numerous in the Southern Ocean. This fall in pH causes a reduction in the number of carbonate ions present in seawater, making the biomineralization process more difficult and more costly in terms of energy. Moreover, planktonic microorganisms, together with some shells or skeletal elements, particularly in larvae and juveniles, could be exposed to dissolution, as seawater becomes corrosive. This could indirectly affect entire sectors of trophic webs, since plankton is at the base of the food chain. One last consequence of changes in pH is related to the physiological balance of organisms. Most are osmoconforming, and so the pH of their internal fluids is linked to that of seawater. If the pH level is modified, will they be able to compensate for the changes induced by acidification?

7.2. The next cold event

The alternation of glacial–interglacial cycles should continue, meaning that in the next few thousand years Earth will enter a new glacial period. A review of past fluctuations suggests that the next glacial period will occur in 15 to 20,000 years [DEC 14]. Such climatic upheaval will be accompanied by major changes in the Southern Ocean, which will undergo a much vaster and more permanent state of glaciation than today. Average temperatures during glacial maxima are around 8°C lower than interglacial temperatures (values based on data from the Vostok and Dôme C ice cores) and ocean surface temperatures drop by around 2.5°C. This reversal could counterbalance the effects of anthropogenic changes, provided that they have not pushed the oceanic system and its thermohaline circulation into a different equilibrium by then. The uncertainty surrounding these eventualities is based on the level of climate upheaval induced by human activities at the time of the new glacial period. If warming is so great that Earth is plunged back into a *greenhouse*-type warm period, like those that occurred in the distant past, then the glacial period will bring conditions back roughly to where they are now. This extreme eventuality is very unlikely because it would imply a World Ocean with an average surface temperature of 25 to 30°C, the result of a very warm atmosphere. The most likely case is that anthropogenic warming will remain within the limits of a 3 to 5°C rise, meaning that the next oscillation will lead to a true glacial event.

During this next glacial period, the large brown algae of the Southern Ocean could extend their distribution area towards the tropics, as they did during the last glacial event [OPP 93]. Species that have benefitted from anthropogenic global warming and settled near the Antarctic will leave again and move northward, disrupting ecosystems once more.

7.3. Drifting continents

In the more distant future, after tens of millions of years, continental drift should lead to even greater isolation of the Antarctic continent. This hypothesis belongs to the domain of geo-fiction, but what would such predictions look like (see Figure 7.4)?

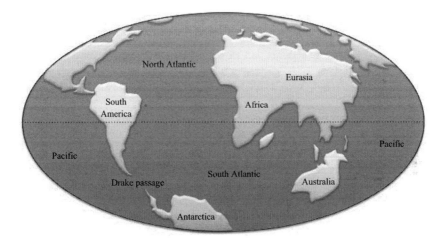

Figure 7.4. *Geography of the future. Earth in 50 million years time. Modified from Scotese (http://www.scotese.com/future.htm)*

Plate tectonics will cause the northward drift of Africa, closing the Mediterranean and opening even more broadly an immense South Atlantic Ocean. Australia will occupy almost the same position as today, but the northwesterly drift of Antarctica will cause a vast ocean to appear between it and Australia. This drift northwestward will preserve almost the same short distance as today with South America, which will in turn experience a strong shift northward. In terms of climate, the Antarctic continent should be in a less polar position and therefore benefit from more clement conditions. But what will the climate be like in such a distant future? We prefer not to run the risk of attempting to predict ecosystems so far into the future.

Appendix

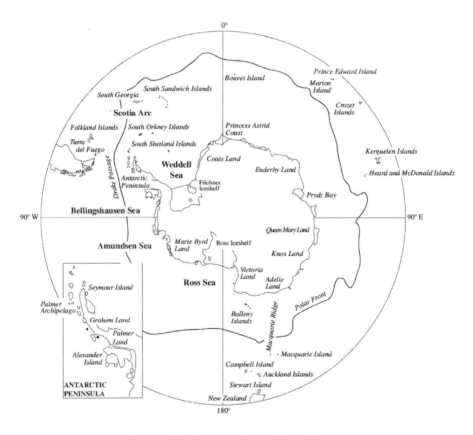

Figure A.1. *Map of the Southern Ocean*

Era	Period	Epoch	Age	
Cenozoic	Quaternary	Pleistocene	-2.59 not detailed	
		Pliocene		-5.33
				-7.25
	Neogene	Miocene	Messinian	-11.63
			Tortonian	-13.82
			Serravalian	-15.97
			Langhian	-20.44
			Burdigalian	-23.03
			Aquitanian	
	Paleogene	Oligocene	Chattian	-28.1
			Rupelian	-33.9
		Eocene	Priabonian	-37.8
			Bartonian	-41.2
			Lutetian	-47.8
			Ypresian	-56
		Paleocene	Thanetian	-59.2
			Selandian	-61.6
			Danian	-66
Mesozoic (2/2)	Cretaceous	Late	Maastrichtian	-72.1
			Campanian	-83.6
			Santonian	-86.3
			Coniacian	-89.8
			Turonian	-93.9
			Cenomanian	-100.5
		Early	Albian	-113
			Aptian	-126.3
			Barremian	-130.8
			Hauterivian	-133.9
			Valanginian	-139.4
			Berriasian	-145

Figure A.2a. *Geological time scale (Cenozoic and second part of the Mesozoic). Ages in millions of years. From [GRA 12]*

Era	Period	Epoch	Age	
Mesozoic (1/2)	Jurassic	Late	Tithonian	-152.1
			Kimmeridgian	-157.3
			Oxfordian	-163.5
		Middle	Callovian	-166.1
			Bathonian	-168.3
			Bajocian	-170.3
			Aalenian	-174.1
		Early	Toarcian	-182.7
			Pliensbachian	-190.8
			Sinemurian	-199.3
			Hettangian	-201.3
	Triassic	Late	Rhetian	-209.5
			Norian	-228.4
			Carnian	-237
		Middle	Ladinian	-241.5
			Anisian	-247.1
		Lower	Olenekian	-250
			Induan	-252.2

Figure A.2b. *Geological time scale (first part of the Mesozoic). Ages in millions of years. From [GRA 12]*

Bibliography

[ALI 08] ALI J.R., AITCHISON, J.C., "Gondwana to Asia: Plate tectonics, paleogeography and the biological connectivity of the Indian sub-continent from the Middle Jurassic through latest Eocene (166-35 Ma)", *Earth-Science Reviews*, no. 88, pp. 145-166, 2008.

[ALL 78] ALLEN J.A., "The geographical distribution of the Mammalia, considered in relation to the principle ontological regions of the Earth, and the laws that govern the distribution of animal life", *Bulletin of the US Geological Survey*, no. 4, pp. 1-376, 1878.

[ARN 74a] ARNAUD P.M., "Les anomalies dimensionnelles des organismes benthiques, antarctiques et subantarctiques", *Téthys*, no. 6, pp. 589-601, 1974.

[ARN 74b] ARNAUD P.M., "Les phénomènes d'incubation chez les invertébrés benthiques", *Téthys*, no. 6, pp. 577-588, 1974.

[ARN 97] ARNTZ W.E., GUTT J., KLAGES, M., "Antarctic marine biodiversity: an overview", in ARNTZ W.E., GUTT J., KLAGES M. (eds), *Antarctic Communities: Species, Structure, and Survival*, Cambridge University Press, Cambridge, pp. 3-13, 1997.

[ARO 07] ARONSON R.B., THATJE S., CLARKE A. *et al.*, "Climate change and invasibility of the Antarctic benthos", *Annual Review of Ecology, Evolution, and Systematics*, no. 38, pp. 129-154, 2007.

[ARO 09] ARONSON R.B., MOODY R.M., IVANY L.C. *et al.*, "Climate change and trophic response of the Antarctic bottom fauna", *PLoS ONE*, no. 4, e4385, 2009.

[ARO 15] ARONSON R.B., FREDERICH M., PRICE R. *et al.*, "Prospects for the return of shell-crushing crabs to Antarctica", *Journal of Biogeography*, no. 42, pp. 1-7, 2015.

[BAR 08] BARNES D.K.A., GRIFFITHS H.J., "Biodiversity and biogeography of southern temperate and polar bryozoans", *Global Ecology and Biogeography*, no. 17, pp. 84-99, 2008.

[BER 96] BERKMAN P.A., PRENTICE M.L., "Pliocene extinction of Antarctic pectinid mollusks", *Science*, no. 271, pp. 1606-1607, 1996.

[BEU 09] BEU A.G., "Before the ice: biogeography of Antarctic Paleogene molluscan faunas", *Palaeogeography, Palaeoclimatology, Palaeoecology*, no. 284, pp. 191-226, 2009.

[BRA 08] BRACEGIRDLE T.J., CONNOLLEY W.M., TURNER J., "Antarctic climate change over the 21st Century", *Journal of Geophysical Research*, no. 113, D03103, 2008.

[BRA 07] BRANDT A., GOODAY A.J., BRIX S.B. *et al.*, "The Southern Ocean deep sea: first insights into biodiversity and biogeography", *Nature*, no. 447, pp. 307–311, 2007.

[BRA 09] BRANDT A., LINSE K., SCHÜLLER M., "Bathymetric distribution patterns of Southern Ocean macrofaunal taxa: Bivalvia, Gastropoda, Isopoda and Polychaeta", *Deep-Sea Research I*, no. 56, pp. 2013–2025, 2009.

[BRI 07] BRIGGS J.C., "Marine biogeography and ecology: invasions and introductions", *Journal of Biogeography*, no. 34, 193-198, 2007.

[BRO 01a] BROCKINGTON S., CLARKE A., CHAPMAN A.L.G., "Seasonality of feeding and nutritional status during the austral winter in the Antarctic sea urchin *Sterechinus neumayeri*", *Marine Biology*, no. 169, pp. 127-138, 2001.

[BRO 01b] BROCKINGTON S., PECK L.S., "Seasonality of respiration and ammonium excretion in the Antarctic echinoid *Sterechinus neumayeri*", *Marine Ecology Progress Series*, no. 219, pp. 159-168, 2001.

[BRO 95] BROWN J.H., *Macroecology*, University of Chicago Press, Chicago, 1995.

[CAM 90] CAMLR, Rapport de la neuvième réunion du comité scientifique pour la conservation de la faune et la flore marines de l'Antarctique, CCAMLR, Hobart, 1990.

[CLA 89] CLARKE A., CRAME J.A., "The origin of the Southern Ocean marine fauna", in CRAME J.A. (ed.), *Origins and Evolution of the Antarctic Biota*, Special Publications of the Geological Society of London, London, no. 47, pp. 253-268, 1989.

[CLA 10] CLARKE A., CRAME J.A., "Evolutionary dynamics at high latitudes: speciation and extinction in polar marine faunas", *Philosophical Transactions of the Royal Society B*, no. 365, pp. 3655–3666, 2010.

[CLA 04] CLARKE A., ARONSON R.B., CRAME J.A. *et al.*, "Evolution and diversity of the benthic fauna of the Southern Ocean continental shelf", *Antarctic Science*, no. 16, pp. 559-568, 2004.

[CLA 05] CLARKE A., BARNES D.K.A., HODGSON D.A., "How isolated is Antarctica?", *Trends in Ecology and Evolution*, no. 20, pp. 1-3, 2005.

[CLA 08] CLARKE A., "Antarctic marine benthic diversity: patterns and processes", *Journal of Experimental Marine Biology and Ecology*, no. 366, pp. 48-55, 2008.

[CRA 93] CRAME J.A., "Bipolar molluscs and their evolutionary implications", *Journal of Biogeography*, no. 20, pp. 145-161, 1993.

[CRA 99] CRAME J.A., "An evolutionary perspective on marine faunal connections between southernmost South America and Antarctica", *Scientia Marina*, no. 63, pp. 1–14, 1999.

[CRA 04] CRAME J.A., "Pattern and process in marine biogeography: a view from the poles", in LOMOLINO M.V., HEANEY L.R. (eds), *Frontiers of Biogeography: New Directions in the Geography of Nature*, Sinauer, Sunderland, pp. 271-291, 2004.

[CUN 95] CUNNINGHAM W.D., DALZIEL I.W.D., LEE T.-Y. *et al.*, "Southernmost South America–Antarctic Peninsula relative plate motions since 84 Ma: implications for the tectonic evolution of the Scotia Arc region", *Journal of Geophysical Research*, no. 100, pp. 8257-8266, 1995.

[CZI 14] CZIKO P.A., DEVRIES A.L., EVANS C.W. *et al.*, "Antifreeze protein-induced superheating of ice inside Antarctic notothenioid fishes inhibits melting during summer warming", *Proceedings National Academy of Sciences*, no. 111, pp. 14583-14588, 2014.

[DAL 13a] DALY M., RACK F., ZOOK R., "*Edwardsiella andrillae*, a new species of sea anemone from Antarctic ice", *PLoS ONE*, no. 8, e83476, 2013.

[DAL 13b] DALZIEL I.W.D., LAWVER L.A., PEARCE J.A. *et al.*, "A potential barrier to deep Antarctic circumpolar flow until the late Miocene?", *Geology*, no. 41, pp. 947-950, 2013.

[DAV 88] DAVENPORT J., "Tenacity of the Antarctic limpet *Nacella concinna*", *Journal of Molluscan Studies*, no. 54, pp. 355-356, 1988.

[DAV 90] DAVID B., MOOI R., "An echinoid that 'gives birth': morphology and systematics of a new Antarctic species, *Urechinus mortenseni* (Echinodermata, Holasteroida)", *Zoomorphology*, no. 110, pp. 75-89, 1990.

[DAV 02] DAVISON W., FRANKLIN C., "The Antarctic nemertean *Parborlasia corrugatus*: an example of an extreme oxyconformer", *Polar Biology*, no. 25, pp. 238-240, 2002.

[DAY 71] DAYTON P.K., ROBILLARD G.A., "Implications of pollution to the McMurdo Sound benthos", *Antarctic Journal*, no. 6, pp. 53-56, 1971.

[DEB 96] DE BROYER C., JAZDZEWSKI K., "Biodiversity of the Southern Ocean: towards a new synthesis for the Amphipoda (Curstacea)", *Bolletino del Museo Civico di Storia Manutrale di Verona*, no. 20, pp. 547-568, 1996.

[DEB 11] DE BROYER C., DANIS B., "How many species in the Southern Ocean? Towards a dynamic inventory of the Antarctic marine species", *Deep-Sea Research II*, no. 58, pp. 5-17, 2011.

[DEB 14] DE BROYER C., KOUBBI P., GRIFFITHS H. *et al.*, *Biogeographic Atlas of the Southern Ocean*, SCAR, Cambridge, 2014.

[DEC 14] DECONINCK J.F., *Paléoclimats. L'enregistrement des variations climatiques*, Vuibert, Paris, 2014.

[DEL 72] DELL R.K., "Antarctic benthos", *Advances in Marine Biology*, no. 10, pp. 1-216, 1972.

[DIA 11] DÍAZ A., FÉRAL J.-P., DAVID B. *et al.*, "Evolutionary pathways among shallow and deep-sea echinoids of the genus *Sterechinus* in the Southern Ocean", *Deep-Sea Research II*, no. 58, pp. 205–211, 2011.

[DOU 14] DOUGLASS L.L., TURNER J., GRANTHAM H.S. et al., "A hierarchical classification of benthic biodiversity and assessment of protected areas in the Southern Ocean", *PLoS ONE*, no. 9, e100551, 2014.

[DOW 12] DOWNEY R.V., GRIFFITHS H.J., LINSE K. et al., "Diversity and distribution patterns in high southern latitude sponges", *PLoS ONE*, no. 7, e41672, 2012.

[DUC 07] DUCKLOW H.W., BAKER K., MARTINSON D.G. et al., "Marine pelagic ecosystems: the west Antarctic Peninsula", *Philosophical Transaction of the Royal Society London*, no. 362, pp. 67-94, 2007.

[DUD 05] DUDICOURT J.C., NERAUDEAU D., NICOLLEAU P. et al., "Une faune remarquable d'échinides marsupiaux dans le Pliocène de Vendée (Ouest de la France)", *Bulletin de la Société Géologique de France*, no. 176, pp. 545-557, 2005.

[EGG 02] EGGINGTON S., SKILBECK C., HOOFD L. et al., "Peripheral oxygen transport in skeletal muscle of Antarctic and sub-Antarctic notothenioid fish", *The Journal of Experimental Biology*, no. 205, pp. 765-779, 2002.

[EAS 00] EASTMAN J.T., MCCUNE A.R., "Fishes on the Antarctic continental shelf: evolution of a marine species flock?", *Journal of Fish Biology*, no. 57, pp. 84-102, 2000.

[EKM 53] EKMAN S., *Zoogeography of the Sea*, Sidgwick and Jackson, London, 1953.

[EXO 04] EXON N.F., KENNETT J.P., MALONE M.J., "Leg 189 synthesis: Cretaceous-Holocene history of the Tasmanian Gateway", in EXON N.F., KENNETT J.P., MALONE M.J. (eds), *Proceedings ODP Scientific Results*, no. 189, pp. 1-37, 2004.

[FLE 96] FLESSA K.W., JABLONSKI D., "The geography of evolutionary turnover: a global analysis of extant bivalves", in JABLONSKI D., ERWIN D.H., LIPPS J.H. (eds), *Evolutionary Paleobiology*, University of Chicago Press, Chicago, pp. 376-397, 1996.

[FOX 15] FOX D., "Discovery: Fish live beneath Antarctica", *Scientific American*, no. 21, January 21, 2015.

[FRE 00] FREDERICH M., SARTORIS F.J., ARNTZ W.E. et al., "Haemolymph Mg^{2+} regulation in decapod crustaceans: physiological correlates and ecological consequences in polar areas", *Journal of Experimental Biology*, no. 203, 1383-1393, 2000.

[FRE 07] FRENOT Y., *Régions polaires: quels enjeux?*, Le Pommier/Cité des sciences et de l'industrie, Paris, 2007.

[GIL 84] GILL T., "The principles of zoogeography", *Proceedings of the Biological Society of Washington*, no. 2, pp. 1-39, 1884.

[GIE 13] GROUPE D'EXPERTS INTERGOUVERNEMENTAL SUR L'EVOLUTION DU CLIMAT (GIEC), Changements climatiques 2013. Les éléments scientifiques. Résumé à l'intention des décideurs, 2013.

[GOE 08] GOEBEL T., WATERS M.R., O'ROURKE D.H., "The late Pleistocene dispersal of modern humans in the Americas", *Science*, no. 319, pp. 1497-1502, 2008.

[GON 10] GONZÁLEZ-WEVAR C.A., NAKANO T., CAÑETE J.I. *et al.*, "Molecular phylogeny and historical biogeography of Nacella (Patellogastropoda: Nacellidae) in the Southern Ocean", *Molecular Phylogenetics and Evolution*, no. 56, pp. 115-124, 2010.

[GRA 06] GRANT S., CONSTABLE A., RAYMOND B. *et al.*, *Bioregionalisation of the Southern Ocean: Report of Experts Workshop*, Hobart, WWF Australia, September 2006.

[GRA 12] GRADSTEIN F.M., OGG J.G., SCHMITZ M.D. *et al.*, *The Geological Time Scale 2012*, Elsevier B.V., Oxford, 2012.

[GRI 09] GRIFFITHS H.J., BARNES D.K.A., LINSE K., "Towards a generalized biogeography of the Southern Ocean benthos", *Journal of Biogeography*, no. 36, pp. 162-177, 2009.

[GRI 10] GRIFFITHS H.J., "Antarctic marine biodiversity – what do we know about the distribution of life in the Southern Ocean?", *PLoS ONE*, no. 5, e11683, 2010.

[GRI 13] GRIFFITHS H.J., WHITTLE R.J., ROBERTS S.J. *et al.*, "Antarctic crabs: Invasion or endurance?", *PLoS ONE* no. 8, e66981, 2013.

[GUT 01] GUTT J., "On the direct impact of ice on marine benthic communities, a review", *Polar Biology*, no. 24, pp. 553-564, 2001.

[GUT 10] GUTT J., HOSIE G., STODDART M., "Marine life in the Antarctic", in MCINTYRE A.D. (ed.), *Life in the World's Oceans: Diversity, Distribution, and Abundance*, Blackwell Publishing Ltd., Oxford, pp. 203-220, 2010.

[GUT 11] GUTT J., BARRATT I., DOMACK E. *et al.*, "Biodiversity change after climate-induced ice-shelf collapse in the Antarctic", *Deep Sea Research II*, no. 58, pp. 74-83, 2011.

[HAW 08] HAWES T.C., WORLAND M.R., BALE J.S., "Freezing in the Antarctic limpet, *Nacella concinna*", *Cryobiology*, no. 61, pp. 128-132, 2010.

[HAW 10] HAWES T.C., WORLAND M.R., BALE J.S., "Physiological constraints on the life cycle and distribution of the Antarctic fairy shrimp *Branchinecta gaini*", *Polar Biology*, no. 31, pp. 1531-1538, 2008.

[HED 69] HEDGPETH J.W., "Introduction to Antarctic zoogeography. Distribution of selected groups of marine invertebrates in water south of 35o.S", in BUSHNELL V.C., HEDGPETH J.W. (eds), *Antarctic Map Folio Series 11*, American Geographical Society, New York, pp. 1-9, 1969.

[HED 71] HEDGPETH J.W., "Perspectives of benthic ecology in Antarctica", in QUAM L. (ed.), *Research in the Antarctic*, American Association for the Advancement of Science, Washington DC, pp. 93-136, 1971.

[HEN 13] HENRY L.V., TORRES J.J., "Metabolism of an Antarctic solitary coral, *Flabellum impensum*", *Journal of Experimental Marine Biology and Ecology*, no. 449, pp. 17-21, 2013.

[HUC 04] HUCKE-GAETE R., OSMAN L.P., MORENO C.A. *et al.*, "Examining natural population growth from near extinction: the case of the Antarctic fur seal at the South Shetlands, Antarctica", *Polar Biology*, no. 27, pp. 304-311, 2004.

[ING 12] INGELS J., VANREUSEL A., BRANDT A. et al., "Possible effects of global environmental changes on Antarctic benthos: a synthesis across five major taxa", *Ecology and Evolution*, no. 2, pp. 453-485, 2012.

[JAB 06] JABLONSKI D., ROY K., VALENTINE J.W., "Out of the tropic: evolutionary dynamics of the latitudinal diversity gradient", *Science*, no. 314, pp. 102-106, 2006.

[JAB 08] JABLONSKI D., "Evolution and the spatial dynamics of biodiversity", *Proceedings of the National Academy of Science USA*, no. 195, pp. 11528-11535, 2008.

[KAI 13] KAISER S., BRANDAO S.N., BRIX S. et al., "Patterns, processes and vulnerability of the Southern Ocean benthos: a decadal leap in knowledge and understanding", *Marine Biology*, no. 160, pp. 2295-2317, 2013.

[KNO 60] KNOX G.A., "Littoral ecology and biogeography of the southern oceans", *Proceedings of the Royal Society of London B*, no. 152, pp. 577-624, 1960.

[KNO 77] KNOX G.A., LOWRY J.K., "A comparison between the benthos of the Southern Ocean and the North Polar Ocean with special reference of the Amphipoda and the Polychaeta", in DUNBA M.J. (ed.), *Polar Ocean*, Arctic Institute of North America, Calgary, pp. 423-462, 1977.

[KNO 07] KNOX G.A., *Biology of the Southern Ocean*, CRC Press, Boca Raton, 2007.

[KOU 10] KOUBBI P., OZOUF-COSTAZ C., GOARANT A. et al., "Estimating the biodiversity of the East Antarctic shelf and oceanic zone for ecoregionalisation: example of the ichthyofauna of the CEAMARC (Collaborative East Antarctic Marine Census) CAML surveys", *Polar Science*, no. 4, pp. 115-133, 2010.

[KRU 09] KRUG A.Z., JABLONSKI D., VALENTINE J.W., "Signature of the end-Cretaceous mass extinction in the modern biota", *Science*, no. 323, pp. 767-771, 2009.

[KRU 10] KRUG A.Z., JABLONSKI D., ROY K. et al., "Differential extinction and the contrasting structure of polar marine faunas", *PLoS ONE*, no. 5, e15362, 2010.

[LAG 09] LAGABRIELLE Y., GODDÉRIS Y., DONNADIEU Y. et al., "The tectonic history of Drake Passage and its possible impacts on global climate", *Earth and Planetary Science Letters*, no. 279, pp. 197-211, 2009.

[LAW 03] LAWVER L.A., GAHAGAN L.M., "Evolution of Cenozoic seaways in the circum-Antarctic region", *Palaeogeography, Palaeoclimatology, Palaeoecology*, no. 198, pp. 11-37, 2003.

[LEB 07] LE BOHEC C., DURANT J.M., GAUTHIER-CLERC M. et al., "King penguin populations threatened by Southern Ocean warming", *Proceedings National Academy of Sciences*, no. 105, pp. 2493-2497, 2007.

[LEC 04] LECOINTRE G., OZOUF-COSTAZ C., "Les poissons antigel de l'océan Austral", *Pour la Science*, no. 320, pp. 48-54, 2004.

[LEC 13] LECOINTRE G., AMEZIANE N., BOISSELIER M.-C. et al., "How operational is the species flock concept? The Antarctic shelf case", *PLoS ONE*, no. 8, e68787, 2013.

[LEE 04] LEE Y.-H., SONG M., LEE S. et al., "Molecular phylogeny and divergence time of the Antarctic sea urchin (*Sterechinus neumayeri*) in relation to the South American sea urchins", *Antarctic Science*, no. 16, pp. 29-36, 2004.

[LEF 12] LEFEBVRE V., DONNADIEU Y., SEPULCHRE P. et al., "Deciphering the role of southern gateways and carbon dioxide on the onset of the Antarctic Circumpolar Current", *Paleoceanography*, no. 27, PA4201, 2012.

[LIN 06] LINSE K., GRIFFITHS H.J., BARNES D.K.A. et al., "Biodiversity and biogeography of Antarctic and sub-Antarctic mollusca", *Deep-Sea Research II*, no. 53, pp. 985-1008, 2006.

[LIV 05] LIVERMORE R., NANKIVELL A., EAGLES G. et al., "Paleogene opening of Drake Passage", *Earth and Planetary Science Letters*, pp. 459-470, 2005.

[LOM 06] LOMOLINO M.V., RIDDLE B.R., BROWN J.H., *Biogeography*, Sunderland, Sinauer Associates, London, 2006.

[LON 07] LONGHURST A.R., *Ecological Geography of the Sea*, Academic Press, San Diego, 2007.

[LÜD 12] LÜDECKE C., SUMMERHAYES C., *The Third Reich in Antarctica. The German Antarctic Expedition 1938-39*, Erskine Press, Huntingdon, 2012.

[LYT 01] LYTHE M.B., VAUGHAN D.G., BEDMAP Consortium, "BEDMAP: a new ice thickness and subglacial topographic model of Antarctica", *Journal of Geophysical Research*, no. 106, pp. 11335-11352, 2001.

[MIS 12] MISRATI N., FINNEY B.P., JORDAN J.W. et al., "Early retreat of the Alaska Peninsula Glacier Complex and the implications for coastal migrations of First Americans", *Quaternary Science Reviews* no. 48, pp. 1-6, 2012.

[MOR 12] MORAN A.L., WOODS H.A., "Why might they be giants? Towards an understanding of polar gigantism", *The Journal of Experimental Biology*, no. 215, pp. 1995-2002, 2012.

[NAI 09] NAISH T., POWELL R., LEVY R. et al., "Obliquity-paced Pliocene West Antarctic ice sheet oscillations", *Nature*, no. 458, pp. 322-329, 2009.

[OBE 11] OBERMÜLLER B.E., MORLEY S.A., CLARK M.S. et al., "Antarctic intertidal limpet ecophysiology: a winter–summer comparison", *Journal of Experimental Marine Biology and Ecology*, no. 403, pp. 39-45, 2011.

[OPP 93] VAN OPPEN M.J.H., OLSEN J.L., STAM W.T. et al., "*Acrosiphonia arcta* (Chlorophyta) and *Desmaretia viridis/willii* (Phaeophyta) are of recent origin", *Marine Biology*, no. 115, pp. 381-386, 1993.

[ORS 95] ORSI A.H., WHITWORTH III T.W., NOWLIN Jr. W.D., "On the meridional extent and fronts of the Antarctic Circumpolar Current", *Deep Sea Research I*, no. 42, pp. 641-673, 1995.

[ORT 96] ORTMANN A.E., *Grundzüge der Marinen Tiergeographie*, Gustav Fischer, Jena, 1896.

[PAR 09] PARK E.T., FERRARI F.D., "Species diversity and distributions of pelagic calanoid copepods from the Southern Ocean", in KUPNIK I., LANG M.A., MILLER S.E. (eds), *Smithsonian at the poles: Contributions to International Polar Year Science*, Smithsonian Institution Scholarly Press, Washington DC, 2009.

[PEA 69] PEARSE J.S., "Slow developing demersal ambryos and larvae of the Antarctic sea star *Odontaster validus*", *Marine Biology*, no. 3, pp. 110-116, 1969.

[PEA 04] PEARSE J.D., LOCKHART S.J., "Reproduction in cold water: paradigm changes in the 20th Century and a role for cidaroid sea urchins", *Deep-sea Research II*, no. 51, pp. 1533-1549, 2004.

[PEA 09] PEARSE J.S., MOOI R., LOCKHART S.J. *et al.*, "Brooding and species diversity in the Southern Ocean: selection of brooders or speciation within brooding clades?", in KRUPNIK I., LANG M.A., MILLER S.E. (eds), *Smithsonian at the Poles: Contributions to International Polar Year Science*, Smithsonian Institution Scholarly Press, Washington DC, 2009.

[PEA 00] PEARSON J.S., PALMER M.R., "Atmospheric carbon dioxide concentrations over the past 60 million years", *Nature*, no. 406, pp. 695-699, 2000.

[PEC 02] PECK L.S., "Ecophysiology of Antarctic marine ectotherms: limits to life", *Polar Biology*, no. 25, pp. 31-40, 2002.

[PEC 06] PECK L.S., CONVEY P., BARNES D.K.A., "Environmental constraints on life histories in Antarctic ecosystems: tempos, timings and predictability", *Biological Review*, no. 81, pp. 75-109, 2006.

[PEC 13] PECK L.S., BROCKINGTON S., "Growth of the Antarctic octocoral *Primnoella scotiae* and predation by the anemone *Dactylanthus antarcticus*", *Deep-Sea Research II*, no. 92, pp. 73-78, 2013.

[PHI 71] PHILIP G.M., FOSTER R.J., "Marsupiate tertiary echinoids from south-eastern Australia and their zoogeographic significance", *Palaeontology*, no. 14, pp. 666-695, 1971.

[PIE 12] PIERRAT B., SAUCÈDE T., FESTEAU A. *et al.*, "Antarctic, sub-Antarctic and cold temperate echinoid database", *Zookeys*, no. 204, pp. 47-52, 2012.

[PIE 13] PIERRAT B., SAUCÈDE T., BRAYARD A. *et al.*, "Comparative biogeography of echinoids, bivalves, and gastropods from the Southern Ocean", *Journal of Biogeography*, no. 40, pp. 1374-1385, 2013.

[PÖR 99] PÖRTNER H.O., PECK L.S., ZIELINSKI S. *et al.*, "Intracellular pH and energy metabolism in the highly stenothermal Antarctic bivalve *Limopsis marionensis* as a function of ambient temperature", *Polar Biology*, no. 22, pp. 17-30, 1999.

[PÖR 06] PÖRTNER H.O., "Climate-dependent evolution of Antarctic ectotherms: an integrative analysis", *Deep-Sea Research II*, no. 53, pp. 1071-1104, 2006.

[POU 96] POULIN E., FÉRAL J.P. , "Why are there so many species of brooding Antarctic echinoids?", *Evolution*, no. 50, pp. 820-830, 1996.

[REG 14] REGUERO M.A., GELFO J.N., LÒPEZ G.M. et al., "Final Gondwana breakup: the Paleogene South American native ungulates and the demise of the South America–Antarctica land connection", *Global and Planetary Change*, no. 123, pp. 400-413, 2014.

[REM 03] REMY F., *l'Antarctique. La mémoire de la Terre vue de l'espace*, CNRS Éditions, Paris, 2003.

[RUD 02] RUDDIMAN W.F., *Earth's Climate. Past and Future*, WH Freeman and Company, New York, 2002.

[SAU 13] SAUCÈDE T., PIERRAT B., BRAYARD A. et al., "Palaeobiogeography of Austral echinoid faunas: a first quantitative approach", in M.J. HAMBREY, P.F. BARKER, P.J. BARRETT et al., (eds), *Antarctic Palaeoenvironments and Earth–Surface Processes*, Special Publications of the Geological Society of London, London, no. 381, pp. 117-127, 2013.

[SCH 14] SCHMIDT K., ATLINSON A., POND D.W. et al., "Feeding and overwintering of Antarctic krill across its major habitats: the role of sea ice cover, water depth, and phytoplancton abundance", *Limnology Oceanography*, no. 59, pp. 17-36, 2014.

[SEE 09] SEEAR P., TARLING G.A., TESCHKE M. et al., "Effects of simulated light regimes on gene expression in Antarctic krill (Euphausia superba Dana)", *Journal of Experimental Marine Biology and Ecology*, no. 381, pp. 57-64, 2009.

[SIE 09] SIEGERT M.J., FLORINDO F., "Antarctic climate evolution", in FLORINDO F., SIEGERT M.J. (eds), *Antarctic Climate Evolution, Developments in Earth and Environmental Sciences*, Elsevier, Amsterdam, 2009.

[SMI 87] SMITH R.I.L., SIMPSON H.W., "Early 19th Century sealer's refuges on Livingstone Island, South Shetlands islands", *British Antarctic Survey*, no. 74, pp. 49-72, 1987.

[SPA 07] SPALDING M.D., FOX H.E., ALLEN G.R. et al., "Marine ecoregions of the world: a bioregionalization of coastal and shelf areas", *Biosciences*, no. 57, pp. 573-83, 2007.

[STE 06] STEPANJANTS S.D, CORTESE G, KRUGLIKOVA S.B et al., "A review of bipolarity concepts: history and examples from Radiolaria and Mdusozoa (Cnidaria)" *Marine Biology Research*, vol. 2, no. 3, pp. 200-241.

[STI 03] STILWELL J.D., "Patterns of biodiversity and faunal rebound following the K-T boundary extinction event in Austral Palaeocene molluscan faunas", *Palaeogeography, Palaeoclimatology, Palaeoecology*, no. 195, pp. 319-356, 2003.

[STI 04] STILWELL J.D., ZINSMEISTER W.J., OLEINIK A.E., "Early Paleocene mollusks of Antarctica: systematics, paleoecology and paleogeographic significance", *Bulletins of American Paleontology*, no. 367, pp. 1-89, 2004.

[STO 90] STOTT L.D., KENNETT J.P., SHACKLETON N.J. et al., "The evolution of Antarctic surface waters during the Paleogene: Inferences from the stable isotope composition of planktonic foraminifera, ODP Leg 113", *Proceedings ODP Scientific results*, no. 113, pp. 849-863, 1990.

[STR 08] STRUGNELL J.M., ROGERS A.D., RODÖHL P.A. et al., "The thermohaline expressway: the Southern Ocean as a centre of origin for deep sea octopuses", *Cladistics*, no. 24, pp. 853-860, 2008.

[TAV 04] TAVARES M., DE MELO G.A.S., "Discovery of the first known benthic invasive species in the Southern Ocean: the North Atlantic spider crab *Hyas araneus* found in the Antarctic Peninsula", *Antarctic Science*, no. 16, pp. 129-131, 2004.

[THA 05] THATJE S., ANGER K., CALCAGNO J.A. et al., "Challenging the cold: crabs reconquer the Antarctic", *Ecology*, no. 86, pp. 619-625, 2005.

[THA 05] THATJE S., HILLENBRAND C.-D., LARTER R., "On the origin of Antarctic marine benthic community structure", *Trends in Ecology and Evolution*, no. 20, pp. 534-540, 2005.

[THO 03] THOUZEAU C., LE MAO Y., FROGET G. et al., "Spheniscins, avian ß-defensins in preserved stomach contents of the king penguin, *Aptenodytes patagonicus*", *The Journal of Biological Chemistry*, no. 278, pp. 51053-51058, 2003.

[TRE 14] TRÉGUER P.J. "The Southern Ocean silica cycle", *Comptes Rendus Geoscience*, no. 346, pp. 279-286, 2014.

[UNE 09] UNESCO, Global Open Oceans and Deep Seabed (GOODS) – biogeographic classification, IOC Technical Series, no. 84, UNESCO-IOC, Paris, 2009.

[WAL 06] WALLER C.L., WORLAND M.R., CONVEY P. et al., "Ecophysiological strategies of Antarctic intertidal invertebrates facing with freezing stress", *Polar Biology*, no. 29, pp. 1077-1083, 2006.

[WAL 08] WALLER R.G., TYLER P.A., SMITH C.R., "Fecundity and embryo development of three Antarctic deep-water scleractinians: *Flabellum thouarsii, F. curvatum* and *F. impensum*", *Deep-Sea Research II*, no. 55, pp. 2527-2534, 2008.

[ZAC 01] ZACHOS J.C., PAGANI M., SLOAN L. et al., "Trends, rhythms, and aberrations in global climate 65 Ma to present", *Science*, no. 292, pp. 686-693, 2001.

[ZAC 08] ZACHOS J.C., DICKENS G.R., ZEEBE R.E., "An early Cenozoic perspective on greenhouse warming and carbon-cycle dynamics", *Nature*, no. 4514, pp. 279-283, 2008.

[ZIN 79] ZINSMEISTER W.J., "Biogeographic significance of the late Mesozoic and early Paleogene molluscan faunas of Seymour Island (Antarctic Peninsula) to the final breakup of Gondwanaland", in GRAY J., BOUCOT A.J. (eds), *Historical Biogeography, Plate Tectonics, and the Changing Environment*, 37th Biological Colloquium of the Oregon State University, University of Oregon Press, Corvallis, OR, pp. 347-355, 1979.

[ZIN 82] ZINSMEISTER W.J., "Late Cretaceous - early Tertiary molluscan biogeography of the southern circum-Pacific", *Journal of Paleontology*, no. 56, pp. 84-102, 1982.

Index

A, B, C

Adelie Land, 3-4, 80
Alexander, 60, 69
algae, 30, 66-67, 77, 101
amphipods, 47, 59, 66, 77-78, 86
Amundsen Sea, 22, 45, 94
Antarctic Circumpolar Current (ACC), 19, 24, 33, 50, 91
Antarctic Divergence, 20, 25-28, 52
Antarctic Peninsula, 3, 7, 20, 24-26, 34, 38, 41, 60, 63, 93-94, 98
Bellingshausen Sea, 22
bipolar, 56-57
bivalve, 45, 48, 51, 55, 60-61, 64-67, 91, 97
brachiopods, 48, 60, 66, 83, 86
brittle stars, 45, 47, 80, 85-86, 91
bryozoans, 47-51, 55-56, 66, 79-80, 100
cetaceans, 11, 88
corals, 48, 64, 85, 100
crabs, 46, 56, 60-61, 65, 95-96
crinoids, 47
Crozet, 2, 19, 31, 51, 55, 96
crustaceans, 45, 47, 55, 59-68, 71-74, 77-83, 86, 95, 100

D, E, F

decapods, 45-46, 59-61, 65-68, 95-96
Drake Passage, 24-25, 31, 33-35, 38-40, 63, 66, 91-92
drift ice, 22
echinoderms, 45, 47-50, 56, 59-60, 65-66, 70, 83, 85, 91
endemism, 49, 53-55, 63
fast ice, 22-23
fish, 11, 36, 45-47, 50, 56, 59-60, 64-68, 71-78, 82, 91, 96
fur seals, 10-11, 73

G, I, K

gastropod, 45-56, 60-66, 77, 83, 86, 91
glacial, 19, 29, 40-41, 48, 55-56, 61, 65, 67-71, 81, 92, 100-101
Gondwana, 33-36, 60, 63-64, 91
greenhouse, 36, 38, 93, 100
iceberg, 4, 22-24, 29, 39-40, 80, 94
icehouse, 38
interglacial, 41, 56, 65, 68-70, 92, 100

isopods, 47, 59, 62-63, 66, 83, 86, 91
Kerguelen, 2-3, 6, 19, 26, 31, 34, 37-38, 50-51, 55, 88-89, 96
King George, 3, 9, 60, 83, 85, 98-99
krill, 11, 29, 46-47, 66, 74, 77, 80-81

L, M, N

lobsters, 60-61, 95
Marion, 2, 3, 9, 51
microalgae, 23, 29, 47, 75, 81
mollusks, 47, 56, 64, 69-74, 79, 82-83, 100
notothenioids, 64, 66, 68, 72, 74, 80

O, P

octopuses, 62
pack ice, 4, 22, 74-77
Pangaea, 56
penguins, 46-47, 64-66, 73, 75, 81, 98-99
Polar Front, 18, 20, 25-29, 38, 47-48, 52, 55, 95
polychaetes, 47, 59, 62
polynyas, 22-23, 74, 77
Prince Edward, 2, 51, 55
protein, 56, 68, 72, 74, 81
province, 50-54, 63-64
pycnogonids, 45, 47, 48, 51, 55, 59, 86, 97

R, S

ray, 46, 60, 96
Ross Sea, 3-4, 7, 9, 41, 69, 75, 78, 86, 95
Scotia Arc, 55, 63, 68, 92
sea cucumbers, 47, 62, 80
sea ice, 7, 15, 19, 21-22, 27, 39, 41, 68, 71, 74-81, 94-95
sea lions, 73
sea urchins, 45-48, 51-55, 67, 71, 80-92
Seymour, 60, 64-67
sharks, 46, 60, 96
South Georgia, 2, 7, 10, 50-52, 96
South Shetland, 3, 20, 67, 96, 98
sponges, 29, 47, 51, 61-62, 78, 80, 84, 86, 91
squid, 47, 74
starfish, 45, 47, 71, 80, 84-86, 91, 96
Sub-Antarctic Front, 26-27

T, W

Tasman Gateway, 33, 35, 39-40
teleosts, 46, 59-60, 65-66, 72, 74
thermohaline circulation, 25, 28, 62, 100
Thermohaline Expressway, 62-63
Weddell Sea, 7, 13, 19-28, 55, 73-75, 81, 86, 94
Weddellian Isthmus, 34, 63
whales, 4, 10, 47, 77
Wilkes Land, 3-4

Printed and bound by CPI Group (UK) Ltd, Croydon, CR0 4YY

11/06/2025

01899189-0008